PRESENTS

# GROW LIKE A PRO

**THE MARIJUANA ADVANTAGE**     **EDITOR: DANA LARSEN**

QUICK AMERICAN

Grow Like a Pro
Copyright 2003 *Cannabis Culture* magazine.


PROJECT EDITORS: Dana Larsen and S. Newhart

ART DIRECTION: Patty Mitropoulos

TYPESETTING & COVER DESIGN: Lightbourne, Inc.

PRINTED IN CHINA: Everbest Printing Company


Authors and Photographers retain rights to their individual works.

WE THANK ALL CONTRIBUTORS TO THIS BOOK: Greg Allen, Pete Brady,
Barge, DMT, Tom Flowers, Hans, Ashera Jones, Dana Larsen,
Tina Lee, Ed Rosenthal, DJ Short, Anastasia V. & Andrew Young.

The material offered in this book is presented as information that should be
available to the public. The Publisher does not advocate breaking the law.
However, we urge readers to support the secure passage of fair marijuana
legislation.

QUICK AMERICAN
A division of Quick Trading Company
Oakland, California
www.quickamerican.com

ISBN: 0-932551-60-2

# CONTENTS

Introduction • by Dana Larsen 5

**From seed to sprout** • by Hans 6

cannabis culture: AUSTRALIA • by Pete Brady 11

GROW Organic • by DJ Short 14

cannabis culture: JAMAICA • by Pete Brady 17

SOIL FOR YOUR SENSI • by DMT 22

The wick method • by Tina Lee 28

photo gallery • BARGE 30

ORGANIC HYDROPONICS • by Hans 32

cannabis culture: AFGHANISTAN • by Pete Brady 36

Aeroponic Supersonics • by Ashera Jones 38

cannabis culture: MOROCCO • by Pete Brady 44

Timing is EVERYTHING • by DJ Short 48

Control your cannabis • by DMT 50

cannabis culture: RUSSIA • by Anastasia V. 52

LET THEM BREATHE! • by DJ Short 55

photo gallery • PETE BRADY 58

POT POTENCY • by DMT 60

Love Your Mother • by DMT 66

cannabis culture: CHINA • by Greg Allen 68

CLONING MADE EASY • by Ed Rosenthal 72

# CANNABIS CULTURE

*by Dana Larsen*
*Editor,* Cannabis Culture *magazine*

WELCOME TO *CANNABIS CULTURE'S* SPECIAL

*GROW LIKE A PRO*, WHICH WILL GIVE YOU ALL THE

INFORMATION AND INSPIRATION YOU NEED TO

PRODUCE WORLD-CLASS MARIJUANA.

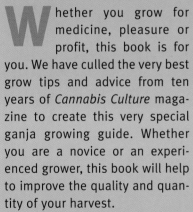

**W**hether you grow for medicine, pleasure or profit, this book is for you. We have culled the very best grow tips and advice from ten years of *Cannabis Culture* magazine to create this very special ganja growing guide. Whether you are a novice or an experienced grower, this book will help to improve the quality and quantity of your harvest.

The cultivation of marijuana is an ancient art that predates recorded history. Since ancient times, this most magical and useful of plants has been providing humanity with clothing, food, shelter and enlightenment. People have been discussing the best cannabis breeding and grow tips for as long as there has been language, maybe longer.

So as you begin your journey of growing this plant, know that you are joining humanity's most ancient fraternity. Remember that the seed you sow comes from over 10,000 generations of careful breeding and selection by ganja growers who loved this plant as much as you.

*Cannabis Culture* magazine is more than just a marijuana information source; we are also an activist publication. Although we provide stories on current cannabis news, events and grow tips, we also work to inspire our readers, to encourage them to help liberate this plant, and to end the tyranny which oppresses pot-people around the globe.

We believe that growing pot is more than just a fun and profitable way to make a living. For we know that this plant can help to open our minds, heal the earth and change our society for the better.

I encourage ganja growers to be brave but cautious, and to know that you are the cornerstone of our culture, preserving and enhancing this sacred plant in your basement, your backyard, or a hidden outdoor locale.

Without those brave enough to keep on growing cannabis, despite prohibition and the threat of imprisonment, there would be no ganja culture. Despite being banned everywhere on Earth, cannabis grows in every city and town in every nation on the globe.

Cannabis is a survivor, finding a place to grow on even the most barren soil and unforgiving climate. So too is the culture which surrounds it. Strong, resilient and determined, we will continue to grow and celebrate this most wonderful of plants until it is free once again, to grow openly and proudly under the wide blue sky.

There is something magical about planting a seed, watching it grow, and reaping the harvest. In every seed there is a bit of freedom — the freedom to sustain yourself.

If you want to grow marijuana, the selection of the seed is one of the most important decisions you will make. Growing good marijuana is an equal combination of genetics and the proper environment. With the right seed in the right environment you will be rewarded with a world-class crop.

# from seed to sprout

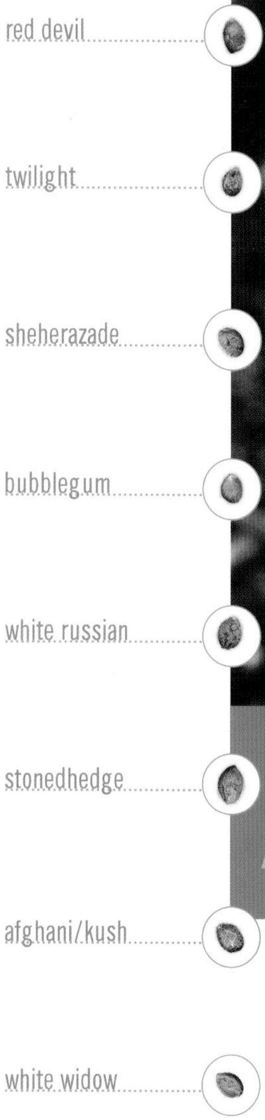

by Hans

Photo: Barge

red devil

twilight

sheherazade

bubblegum

white russian

stonedhedge

afghani/kush

white widow

## HOW TO GERMINATE YOUR BABIES AND START YOUR GARDEN GROWING.

**Y**ou may find a good viable seed in your bag, but seeds acquired in this manner are likely to have been produced by hermaphrodites, and carry that trait.

Quality seeds may be purchased over the counter in Amsterdam and Canada. They can also be purchased from a number of seed companies in Amsterdam and Canada via mail order. A rich and varied selection of seed strains from reputable breeders are available through Marc Emery's seedbank, which advertises in every issue of *Cannabis Culture* magazine. Emery's catalog also lists some basic security techniques to follow when ordering seeds. This is good advice to follow regardless of which company you might purchase from.

Marijuana seeds come in a variety of sizes and colors. They vary in

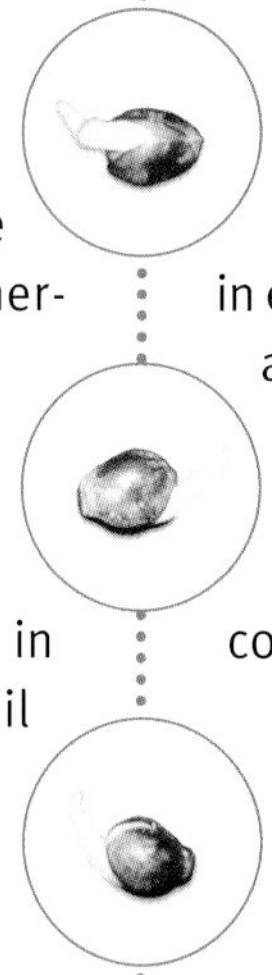

*Germinated seed photos: Scott*

size, from a range of about 1/8 inch to 1/4 inch. Their shapes are round to slightly oblong. Seeds range in color from brown to grey to almost black. Some seeds are plain while others have irregular shaped, different colored lines, such as "lightning" or "tiger" stripes.

A good sign that a seed is viable and healthy will be the fresh "waxy" sheen it displays on its outer coat. A white or light green colored seed is usually the sign of an immature seed. If a seed is black or dark brown and has a dull "non-waxy" appearance, the seed may be dead.

Make a quick physical examination of each seed with the use of a magnifying glass, checking for cracks, molds and other imperfections, to eliminate any "bad seeds."

## Germination

Regardless of the growing method you will ultimately use, the first thing you will need to do is to germinate your seeds. There are a few standard ways this is done.

### PAPER TOWELS

One common and easy way to germinate seeds is to place the seeds in between the layers of a wet paper towel or a cloth such as a clean washrag. Tap water can work fine, but it is better to use store-bought distilled water for all stages of the germination process.

The damp, but not soaked, paper towel or washcloth is placed into a suitably sized plastic bag or simply covered with plastic wrap, and then placed onto a glass plate and kept in a warm, dark place.

The temperature should be maintained between 75-85°F (24-29°C). Horticultural heating pads are excellent for this, however

From top to bottom:

- carefully adding soil to seed pots
- seedlings sprouting in lava rock
- flat trays of happy seedlings

*Photo: Hans*

when using paper towels and washcloths you must check every day to make sure they do not dry out. Your germination medium should *never* dry out, as this will kill the fragile embryonic main root. However, also be sure to not overwater. The seeds can drown if there is standing water in the bag or plate. The key words here are "constantly moist."

It is a good idea to place some type of B1 additive in the water used to germinate the seeds. There are a number of types of this vitamin supplement sold commercially. There is Ortho Up-Start, Super Thrive, Hormex, Power Thrive and a number of other additives which have this vitamin supplement. The additive should only be placed in the water the first time you moisten the towel, and should not be added with each subsequent watering.

The moisture content of the towel must be checked regularly. It usually takes anywhere from three to twelve days for most seeds to germinate. Be patient, as some seeds can take even longer to sprout.

As the moisture enters the softened shell, the embryo within begins to grow and swell. Once the shell has broken open, a single embryonic root will appear. Next the first rounded set of leaves will emerge and throw off the split shell. Once this happens you are ready to carefully move and transplant the tiny sprouts into the medium of choice. This should be done carefully so as not to damage the fragile sprouts.

### SOIL

Some people prefer to germinate their seeds in the same medium in which the mature plant will be grown. However, I would not suggest germinating seeds directly in outdoor soil, even if the final destination of the plants will be outside. Your precious seeds will have a higher survival rate and be healthier if germinated indoors.

There are many different soil mixtures which facilitate seed germination. One of the best I have found starts with a gallon each of

door will be helpful. If the materials are moistened with the distilled water before mixing they will produce less dust as you work.

The pH of any medium should be kept between pH 6 (slightly alkaline) and pH 7 (neutral). Marijuana's ideal pH is 7. This is a neutral pH – it means your pH is perfectly balanced. A handheld pH meter will give an accurate

to the top. The mixture is lightly pressed into the cup. The medium should be watered until water drips out the bottom of the container. If peat moss cups are used, the peat cup should be moistened with tepid water.

I like to place the cups in 12 inch x 24 inch germinating trays or flats. Each flat will hold a certain amount of containers. The flats

This seedling was sprouted in soil placed within a rockwool cube.

Seedlings are often started under fluorescent lighting to avoid burning.

A male (left) and female plant enjoying some foreplay before getting jiggy with it.

potting soil and coarse horticultural perlite. To this mix I add two heaping tablespoons of horticultural hydrated lime. The lime helps to balance the pH and will add secondary nutrients like calcium and magnesium. I also like to add a half gallon of some other nitrogen-bearing organic, such as pasteurized cow manure, pasteurized worm castings, or pasteurized bat guano. The organic ingredients will supply nutrients and their fiber will help texture the man-made soil.

The materials are mixed together thoroughly. A mask should be worn whenever these ingredients are mixed. If they are mixed indoors, a fan blowing over the work area and out a window or

reading of most mediums you will use. Commercial soils are usually never alkaline; if anything they are closer to neutral or acid. This means that you may sometimes find that your pH is slightly lower than it should be. If this is the case, put a tablespoon of horticultural hydrated lime into a gallon of water, and pour it over the medium. The next day check the pH and see how it has changed. Add more solution if needed, and check the pH of your medium every few days.

Seeds germinated in soil should be germinated in smaller two- to three-inch containers. The most common of these are the peat moss cups or plastic cups. The soil mixture is placed in the cup almost

will help the containers retain heat, and can be placed directly on top of heating pads. These trays are also sold with plastic see-through tops, which will create a greenhouse atmosphere and raise the humidity.

Non-heated plastic cups will hold heat better than non-heated peat moss cups. The evaporation of moisture from the peat moss will lower the temperature of the sides of the peat cups. For this reason either a heating pad or some other assurance of temperature control should be used. The seedlings require an 75-85°F (24-29°C) ambient temperature.

A small device such as a chopstick is used to make a hole approximately 1/2-inch deep in the center of the pre-moistened medium. The seed is placed in the hole, pointed end down. Cover the seed with soil using your finger. The container of medium is then lightly watered with a solution of a vitamin B1 supplement.

## ROCKWOOL AND OTHERS

Some growers will want to germinate their seeds in rockwool cubes, coconut fiber cubes, lava rock or Oasis felt-type cubes. This is easily done. These items are all porous materials, which have the ability to hold water for a long period of time.

Rockwool is a heated and spun material, which has some environmental concerns. Rockwool cubes should be pH balanced before use. Coconut fiber will biodegrade faster than rockwool. The Oasis cubes are made of a pH neutral, porous felt-type material, and are one of my favorite mediums for germination. They may be purchased as 12 inch x 24 inch slabs, which fit perfectly into similar-size rooting flats.

Regardless of whichever of these mediums is used for germination, the process is the same. First place the medium in the flats and moisten them to saturation point. They are watered in the flats and if needed a hole is punched in the center top of each cube. A seed is placed point end down. The cubes are watered a second time, this time with the vitamin B1 root stimulant, after which any extra or standing water is drained off.

Plastic covers may or may not be needed to maintain additional humidity. The flats are placed on heating mats, and placed in a dark place. Water as required. The cubes are kept moist but there

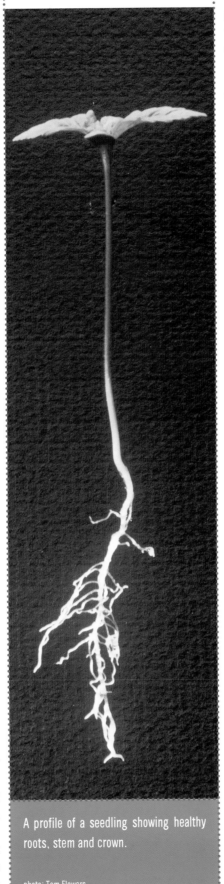

A profile of a seedling showing healthy roots, stem and crown.

photo: Tom Flowers

should be no standing water in the trays. Always pour off any excess water after watering.

## The light of life

Once your seed has germinated it is a seedling. It is now very fragile. The lower end of the young sprout is a single main root. This root will aim itself downward as it follows gravity. On the opposite end of the sprout are two rounded leaves called the *cotyledons*. If the seedlings were germinated in a paper towel they should be immediately moved into some kind of larger container, such as a two- or three-inch container of man-made soil, or a similar-size medium.

Once the seedlings have sprouted they are ready to begin photosynthesis. This means they will need some light. The seedlings may be placed under metal halide or fluorescent light. The lights are left on for 18 to 24 hours a day – I prefer 18 hours.

Metal halide light is powerful, and you must be careful when using them with seedlings. A stationary 400-watt light should be placed with the bottom of the light about three feet above the top of the seedlings. The seedlings should be monitored every few minutes for the first few hours, for damage by burning. The light may be lowered six inches a day, until the light is about two feet over the top of the seedlings. Seedlings should be carefully monitored after each light lowering. Any indication of burning means to immediately raise the lights to their previous setting.

I like to start my seedlings under two-tube, four-foot, 40-watt fluorescent lights. Horticultural grow lights may be used, or a combination of one Cool-White tube (blue spectrum) and one Warm-White or

Soft-White (red spectrum). These lights may be placed with the bottom of the tubes about two inches from the tops of the seedlings. As the seedlings grow, the lights are raised equally, keeping the tubes close to the top of the seedlings. If more than one of these fluorescent lights is being used, they are placed parallel to each other, about six inches apart.

## Rapid Growth

Once the seedlings become exposed to a light source they begin to grow at an accelerated rate. The single embryonic root begins to form lateral rootlets. These lengthening rootlets absorb more nutrients and water each day. These young roots may be damaged by a lack or excess of water, as well as exposure to light, heat, cold or rough treatment.

The roots like access to air to breathe, so the plant is watered, and then the medium is allowed to almost dry out. The medium is allowed to dry to the point where it is not completely dry; it is still just slightly moist. This drying of the medium allows air to be drawn into the medium. The medium is never allowed to dry out completely! Just before the medium can dry completely, the medium is watered again and allowed to almost dry again. This process is repeated.

The rounded cotyledon leaves are soon joined by a pair of the more recognizable serrated leaves. The new plant aims upward as it reaches for the light. Upward growth will continue as more leaves are created. As each new set of leaves appear on the plant, the

space between the sets of leaves will lengthen. The spaces between the sets of leaves are called the internodes. The growth areas where these leaf pairs and lateral growth appear are called the nodes.

The stem of the plant carries nutrients and water up the plant, where they are used on the way and also to create new foliage. If the seedlings have an oscillating fan blowing on them, the plant will produce internal cellulose to create larger and thicker stems as the plant attempts to compensate for the additional environmental interaction. A thicker stem now helps support bigger buds later. A light misting of water a few times a day during lights-on time will help to keep the surfaces of the leaves clean. This will help the leaves with their processes of inhaling and exhaling, as well as photosynthesizing.

The relative humidity should be kept around 50%, and the ambient temperature should be lowered to around 73-78°F (23-26°C) This lower temperature is more conducive to the vegetative state of the larger growing plant. The seedlings may be fed a standard full-spectrum vegetative feeding of any organic or chemical food. Most packaging on vegetative food will indicate the proper amount of the food to give to the seedlings, but remember its always better to give too little than too much.

All feedings must be done with a full-spectrum fertilizer containing all twelve essential elements. Regular fertilizers contain only the three

primary nutrients: Nitrogen (N), Phosphorus (P) and Potassium (K). A full-spectrum fertilizer contains the proper proportions of all the secondary elements and trace elements. Remember to use full spectrum only.

After the seedlings are about two weeks old they should have developed some nice roots and created some lush foliage. The seedling's roots will have used up most of the space they have in the smaller containers, and may already be overrunning them. It will be time to move and transplant them to larger living quarters. If the seedlings are still under fluorescent lights, now is the time to place them under the metal halide.

Congratulations, your plants have survived the most difficult and dangerous part of their lives! With some strains you can start them to flowering almost immediately, or you might prefer to grow them larger first, or to use these plants as clone mothers. Whatever your preference, you are now well on the road to growing yourself some excellent marijuana. 🌿

• Hans is the author of *Sea of Green* and *Organic Hydroponics*, and also a series of cultivation and pot-cooking videos.

# Cannabis Culture:
# Australia!

## HOW GROWERS DOWN UNDER PRODUCE AWARD-WINNING POT UNDER THE WIDE AUSSIE SKY.

## STORY AND PHOTOS BY PETE BRADY

*Time to harvest these fine colas!*

*Buds beginning to form in the land down under.*

*Fine Sativa, standing proud and tall in the Aussie Sun.*

Australia is like another planet. Its geography is impressive. The island continent, bordered by the Pacific Ocean and the Indian Ocean, and approximately as big as the United States, has unending expanses of foreboding desert, a torpid tropical northern region, snowy mountains and lovely greenbelts typified by the lush landscape where Nimbin is situated.

Australia presents a variety of exciting opportunities for cannabis growers. It is "down under" in the southern hemisphere, so Australians enjoy summery Christmas at the beach in December and cool winter days in July.

The country is so vast and uncharted—with 550 national parks, low population density and a still-pervasive agrarian tradition—that most of the marijuana Australians consume is grown in their own country outdoors.

Aussie growers in areas with appropriate seasonal temperature variations start seedlings indoors in October and transplant them outside into Aussie springtime a few weeks later. They look forward to harvests beginning in March and continuing through May. Other growers are able to grow outdoors year round, reaping two to four harvests annually.

Indoor cannabis cultivation is increasingly popular, especially among city dwellers, but ample sunshine, fresh air and freedom from effective aerial surveillance are usually available to Aussie growers in most of the country's fine microclimates.

Outdoor marijuana is different from indoor marijuana. In general, Australian outdoor cultivators work with mostly-Sativa varieties characterized by long thin leaves, extended growing seasons, long internodes and buds, tallness, sturdy stalks and immune systems resistant to wind, intense sun and pests.

Unless Sativas are altered by pruning, training or other interventions, they grow to become Christmas-tree-shaped plants that average 15 feet high, with large lower branches that often create a 3- to 5-foot-wide profile.

Sativas are slow growing plants; some flower intermittently if at all. Others flower when they have reached a certain height or age, or when they have been triggered by daylength or seasonal variations.

Important to the cannabis consumer is the high the Australian Sativa produces. In general a Sativa high is characterized by mental stimulation, mild hallucinations, euphoria, creativity, increased heart rate, anxiety and an absence of "burnout" or sleepiness.

Today's Australian Sativas are probably descendants of seeds imported in the 1970s from Thailand, Southeast Asia, the Pacific Islands and Africa. Aussie growers have also imported pure Indicas and Indica-dominant hybrids such as Haze, Northern Lights, Berry, Skunk and Ice. These non-native Indicas are grown pure or combined with tropical genetics to form potent, adaptable crosses. Although these varieties are initially best suited to indoor grows, Aussies have found ways to keep them stable and adapt them to down under conditions using greenhouses, shadecloths, flower-forcing, mist systems and other techniques.

Key problems with northern hemisphere weed grown in Australia are day length mismatches, inability to handle high temperatures, sun intensity, dehydration and susceptibility to mold.

## GETTING UP DOWN UNDER

The obvious cannatourist destination is the famed cannabis town of Nimbin, located three hours by air from Sydney and approximately one hour by car from the sparkling New Age coastal city of Byron Bay.

Nimbin is a social experiment and ganja trading capital set amongst rainforests and mountain scenery. For many years, the noble band of hippies who settled around Nimbin in the 1970s have held an annual "Mardi Grass" harvest festival every May. During the festival, thousands of people swarm the little village to watch parades featuring "pot fairies" and buses that have huge fake joints on top of them.

Mardi Grass visitors also enjoy the "Hemp Olympics." Contestants compete in various sports, including "fastest joint rolling," "fanciest joint rolling," and "the bong throw." For super-athletes, the "Grower's Triathlon" duplicates some of the adventures of outdoor growers. Competitors must hoist huge sacks of fertilizer and lug big buckets of water through brambles to compete in the event.

Evening Mardi Grass entertainment includes raves, cannabis contests, live music and just chilling out at the many trendy cannabis cafes that line the village's quaint main street.

It's very easy to procure cannabis in Nimbin. Aggressive street vendors, some of whom are junkies or alcoholics, have an annoying habit of asking visitors if they want weed, and acting rudely if the answer is no.

Sometimes, you can pick up Aussie indoor weed, traditional hashish, or even water hashish. Ask to smell, feel and take a hit of weed before you buy it- some Nimbin dealers aren't sure how to properly dry their "mull." Prices are usually lowest in April, May and June when growers are bringing in their biggest outdoor harvests, but you can almost always find at least some good smoke in Nimbin year-round.

The coastal city of Byron Bay is also a good place to score weed and enjoy it. Byron has miles of beautiful beaches, framed by cliffs. Several miles of coastline are protected by national park designation. The park coastline is uncrowded, so you can toke up, relax, make love and tan naked without too much fear of being discovered.

Byron is also one of the largest New Age communities in the world, with more healers, float tanks, massage therapists and personal growth counselors per acre than you'll find most anywhere else. It's fun to get high and get pampered in a spa or on a massage table.

The weed scene in Byron is not quite as out in the open as in Nimbin, but if you walk the main beach and in the downtown's charming shopping core, you can usually find a dealer or be hit on by one.

Remember though, marijuana is very illegal in most parts of Australia, and police have been known to pull some dirty tricks around Nimbin and Byron. These two places are far better for pot people than big cities like Sydney, but be careful if you intend to travel inside Australia with marijuana, especially if you are traveling by air. 🍃

*Phat buds, watch out for wombats!*

*Aussie pot: love it or leaf it!*

# GROW Organic

*If you treat your plants well, they will love and reward you.*

Photo: Barge

•STORY BY DJ SHORT

**IF YOU WANT THE MOST FRAGRANT, DELICIOUS POT ON THE PLANET, THEN USE ORGANIC NUTRIENTS AND FLUSH YOUR BUDS!**

If you want to grow some of the finest herb on the planet, then the True Fragrant varieties of cannabis, such as Blueberry, Blue Velvet and Flo, are an excellent place to start. I speak from experience, as I am the goddess-father of these fine products. The last time I used any seed stock outside of my own was in 1982.

I think that my breeding successes are primarily due to a very discerning palate and sense of smell. A strong and pleasing odor is the dominant feature expressed in the True Fragrant varieties. But you must remember that subtle and subjective characteristics such as "fragrance" and "bouquet" are dependent upon their environment as much as their genetics. I cannot emphasize enough the fact that it takes the purest of environments to grow the purest of herbs.

## BIO VS HYDRO

The purpose of this article is to help guide you in understanding the basic needs of these and other fragrant varieties, and how to best maintain their uniqueness, originality and quality. The key word to this understanding is "organic," or what the

It is very important to give the plant only pure water during the end of the bud cycle. I call this the "rinse and flush" cycle.

The fan leaves should turn yellow during the bud cycle, which shows that the plant is using up all available nutrients.

Photo: Barge

Europeans like to call "bio," methods of production (as opposed to chemical and most hydroponic methods). Simply put, there is no real substitute for the complex relationship of plants and organic soil.

There are those in the hydroponic industry who will argue that certain hydroponic methods are nearly organic and very productive. I don't disagree. However, the main focus of the hydroponic industry is that of production, or quantity, whereas my focus is on quality.

Granted, there are situations where a hydroponic system may be superior to an organic one, especially when the grower wants only one crop and the absolutely highest yield. Sadly, the fact is also that many people simply cannot tell the difference between hydro and organic products, or they simply don't care.

The quality of the hydroponic product may be increased greatly by employing the simple "two week flush" method prior to harvest. This means that only pure water, with no additives or nutrients, be given to the plant for two weeks prior to harvest. This will only slightly decrease production, while greatly increasing the quality of the finished product.

## POTENCY RATIOS

I have found that generally the potency of a given variety of cannabis has to do with the ratio of glandular secreted resins, compared to the overall fiber production of the plant. A higher ratio of resin to fiber generally indicates the superior quality and chemical composition of the resin, and the greater potency of the product. Therefore, in order to maintain potency while increasing

production, this ratio must be maintained. It has been my experience that the more one increases the fiber production and overall size of a given plant, the more one decreases this ratio and, therefore, decreases potency.

This quality/quantity ratio is much less of a concern to the grower who is producing in the great outdoors. I can honestly say from experience that all of the "True Fragrant" varieties are major producers when grown in their particular "sweet spot." Blueberry and Flo have both reached 500 grams per plant, multi-harvested between October 1 and November 7, grown near the 45th parallel in the Pacific Northwest. These plants lost little of their overall appeal despite the increase in production. However, the product of the smaller plants still tended to be more desirable than the larger ones in the outdoor environment.

Someday, when we are allowed to properly produce herb in the great outdoors, we will once again see and experience some of the truly finest examples the planet has to offer. These "fine herbs" come from very specific geographic locations which I refer to as "sweet spots." Certain examples would be: The Northern Californian-Southern Oregon coastal regions; the highland Michoacan, Guerrero, Oaxaca and Chiapas regions of Mexico; highland and valley Colombia; Thailand; the islands of Hawaii; Nepal; parts of Afghanistan; and the Hindu Kush, to name but a few. It is in these "sweet spots" that the most favorable and specifically desirable characteristics are acclimated. Selective inbreeding hardens the desirable characteristics and gives us specific, varied strains. I am very curious to see and experience exactly what our years and multi-generations of indoor breeding are going to produce when returned to these great outdoor "sweet spots."

## INDOOR ORGANICS

Indoor environments are extremely limited in comparison to the great outdoors. The outdoors is a complete and complex system, balanced by many various circumstances. It is sometimes difficult enough to help provide and maintain the proper balances organically in an outdoor garden. Yet although properly providing and maintaining an organic environment indoors is truly a challenge to face, it is not impossible.

Airborne, soil-borne and water-borne pests, fungus, mold, algae and bacteria are just a few of the organisms that can attack a crop and seriously weaken production. It is often too easy to treat these maladies with simple applications of toxic chemicals, and a bit more difficult to solve the problem in a

clean and organic way. Yet here are a variety of adequate organic pesticides and fungicides on the market today. There are also living organisms such as specific predator insects and nematodes. If you feel you must use a commercial chemical product, try to find the least toxic one available for the purpose, and use sparingly. Never apply anything toxic to your plants once they're in the budding cycle.

Another factor to consider is what to use as vitalizers and fertilizers. The bulk of commercial fertilizers and vitalizers (along with most commercial pesticides, herbicides and fungicides) are synthesized from petrochemical by-products and are not truly natural products. Worms, seaweed, bat and bird guano, fish, green manures and most of their by-products are examples of naturally produced substances that provide plenty of good, clean nutrients to the plant. There are now many specific products suited for the indoor organic gardener. Consult your local or

favorite organic garden center for more detail.

## FLUSH YOUR BUDS!

The most important, and perhaps the most simple, aspect to consider involves the last two to three weeks of the bud cycle – the last weeks of the plant's life prior to harvest. It is during this time that absolutely NO additives other than pure water be given to the plant. This is especially important if you have been using chemical fertilizers.

This is the time when the bulk of the final, "useable" part of the plant is produced. As you may well already know, there are over four hundred separate chemicals associated with cannabis and her effects. It is during the final bud-building stage that most of these chemicals are produced. Thus, it is very important to give the plant as much pure water as possible during this crucial period. I like to remember it as the "rinse and flush" cycle. Simply remember to give the plants only water for the last weeks in order to rinse and flush them clean. This is to purge unwanted impurities from the plant.

Pot that has been fertilized right up to harvest is harsh to smoke; sometimes the joint will even sizzle and pop as unmetabolized fertilizer salts combust. Unflushed pot leaves black ash, is hard to keep lit and burns your throat. Pot which has been organically grown and properly flushed is more flavorful and fragrant, burns easily, leaves grey ash, is easier on the throat and is much more pleasurable to smoke. ☙

Photo: DJ Short

# Cannabis Culture:
# JAMAICA

## by Pete Brady

Photos by Pete Brady

The road from the airport in Montego Bay to Kingston was built so long ago that it is totally inappropriate for motor vehicles. I sat terrified in my passenger seat as we sped past uniformed schoolchildren barely an inch away from our bumper, or overtook a truck crowded with screeching chickens on a mountain road so narrow that my door scraped the hedgerow on the side of the road.

My adept driver, other drivers, bicyclists and pedestrians all managed to share the winding road cheerfully and stoically, without resorting to gestures or invectives.

The scenery was spectacular but unsettling—verdant forests, gurgling rivers, waterfalls, hibiscus flowers, scintillating ocean and beaches—littered by shanties, beggars, cripples, crumbling houses, goats, garbage, defecation, sheep, cows, packs of dogs, police with machine guns, cadavers.

My driver fought through the streets of Kingston, the Jamaican capital with its 700,000 population crammed between an ideal harbor and the coffee-famous Blue Mountains that rise behind the city. I had expected an island

Photo by Ed Rosenthal

similar to Hawaii, but instead found myself soaked in sweat in the heart of the Third World.

## JAMAICAN HISTORY

Jamaica has become famous for its volatile political system, uninhibited tourist beaches, Rastafarianism, reggae music and of course, its potent ganja. What you may not know is how Jamaica came to be the place it is today.

Like most North Americans, I've always thought that Jamaica unconditionally welcomed cannabis and cannabis culture. The fame of reggae superstars like Bob Marley and Peter Tosh enmeshed in ganja-shrouded Rastafari religion gave Jamaica a reputation as a country that enthusiastically embraces its ganja roots.

But Jamaica has a considerably more complicated relationship with ganja than one would expect. Cannabis is not native to Jamaica. It was most likely introduced to the island by African slaves and East Indians in the 1700s and 1800s. Its use was associated with working class people, influenced by Hindu beliefs, and a source of increasing worry for the British ruling class who saw ganja as part of a dangerous, rapidly growing black consciousness movement.

Jamaica was brutally colonized by the Spanish in the early 1500s and then stolen by Britain in 1655. Under British rule, the island became a center for Caribbean slave trading. Slaves were forced to build massive sugar cane plantations and work them, a daunting task in the region's blazing sun and tropical humidity. Rocked by slave revolts, the British Parliament abolished slavery in 1838, but free blacks were still

Photo by Pete Brady

victimized by their owners and institutionalized discrimination. They staged another uprising in 1865, fomented by Jamaican folk hero Paul Bogle. The British crushed the uprising in an especially brutal fashion but resistance to vestiges of colonial rule continued, even after the island won independence from the fading British Empire in 1962.

Jamaica's first Prime Minister after independence, Jamaica Labour Party (JLP) leader Alexander Bustamante, was supposedly a handpicked lackey of the British regime. He took office pledging to fight against ganja.

The JLP used the marijuana laws as tools for oppression throughout the 1960s. The war against ganja escalated—police brutality, helicopters, severe prison sentences, reefer madness propaganda and other fascist tactics became commonplace. The People's National Party (PNP) protested JLP's ganja policies and tactics, while Rastafari and Marxist factions gained stature by challenging the imprint of colonialism on the island's socioeconomic and cultural hierarchy.

Then, in 1972, a landmark scientific research study, "Ganja in Jamaica," concluded that marijuana posed no harm to individuals or society. Following the publication of this study, Jamaican officials removed mandatory minimum penalties for ganja use and possession.

The United States continued to fund extensive anti-marijuana campaigns that subsidize the employment of many Jamaicans who work as drug counselors, urine testers, bureaucrats, or for police agencies and private organizations. Attempts to relax ganja laws have been met with threats from the U.S.

*Photo by Pete Brady*

DEA and State Department.

U.S. involvement in Jamaica's internal affairs has angered many Jamaicans. But the island's anti-drug forces aren't sucking off only the American tit: the European Union's Development Fund has given the equivalent of 50 million Jamaican dollars to the country's Integrated Drug Abuse Prevention Project.

Throwing off the yoke of their oppressors, however, Jamaican politicians and officials announced in 2003 that they intended to legalize possession of small amounts of marijuana. Privately, Jamaican officials also indicated that they intended to discontinue most ganja plantation eradication efforts.

"Ganja is not a big problem for Jamaicans," one official said.

Photo by Ed Rosenthal

North Americans come to play. That said, ganja can be purchased from many variety stores and roadside stands throughout the island. They do not overtly advertise ganja sales; they sell candies, coconuts, fruit and other legal goodies. Some of the places I visited were in Kingston, some in a rural area near Negril. Ganja was available in three of the six places I visited. All I had to do was ask.

Just because ganja is relatively easy to come by doesn't excuse stupid behavior, which could lead to personal endangerment or a stint in jail, regardless of what the law says about marijuana possession.

Taxi drivers, Rastafarians and street vendors seem to be safe sources for ganja. Some hotels and resorts advertise themselves as "ganja-friendly," but I visited

"Jamaica's problems are caused by economic and geopolitical policies of the U.S., along with U.S. guns, and cocaine."

## GANJA FOR TOURISTS

Cash can get you anything you want in Jamaica, and ganja is plentiful, but never forget that money transactions in foreign countries, especially where an outsider is purchasing contraband or an illegal service, are opportunities for disaster as well as pleasure.

Most tourists familiar with Jamaica will tell you that the Negril region is the ganja-friendliest. It's close to Montego Bay airport, and home to mega-expensive resorts where Europeans and

Photo by Pete Brady

Photo by Pete Brady

a couple of them and found them dangerous and ghettoish. In resort hotel areas, it is reasonably safe to ask a bartender, taxi driver, waiter or other service provider about procuring ganja. A mellow polite attitude and a healthy dose of common sense helps. And remember: anyone who helps procure anything for you expects payment for that service.

Intuitional and rational antenna should be highly attuned for the rip-off vibe. Some really bad ideas include: going to any drug transaction away from your hotel area and giving your money to somebody who promises to come back with ganja. Likewise, do not go into the back of a store, vending stand or house; make the vendor bring the herb to you. Do

not go alone. Do not buy pre-rolled joints. Beware of street harassment, and be aware that the island's economic situation has deteriorated drastically in the last five years, to the point that even veteran Jamaica visitors are wondering if they should avoid the island until conditions for tourists improve.

Some Jamaicans may be suspicious of you, thinking you are a cop. If you get a bad feeling about somebody, or if the transaction gets tense or weird, get the hell out. Better to find herb somewhere else or even not at all than to get robbed or beaten.

What is the herb like? Overall the herb was surprisingly fresh.

Freshness can be determined by squeezing the bag; if the herb feels dry and brittle, it's probably no good. If it is springy, light reddish-brown or green, has visible resin glands and has been stored in an ice chest (most roadside stands and small stores have no electricity), it is likely to be quality pot.

Savor the difference of outdoor grown bud. Although some tourists told of their experiences with schwagg, most of the herb I saw had few seeds or stems, and had no signs of mold or other pollutants. It was tasty, potent and had a beautiful aroma.

The prices are good, the stone is tropical and the ganja season is year round. Especially if you're a U.S. citizen, don't be tempted to bring any of this inexpensive medicine back home. Almost every flight from Jamaica is greeted by drug war goons and canines. So leave the Jamaican bud behind and bring home photos, Blue Mountain coffee and memories. 🌿

Photo by Ed Rosenthal

The best quality bud will always be produced organically in soil. Hydroponic systems with synthetic nutrient formulas may produce better yields, but they will never touch the quality of the best soil-grown organic pot.

Aside from the quality issue, there are many reasons to choose soil over hydro. Soil gardens pollute much less than hydro ones due to the method of fertilizer application – there are no reservoirs to dump down the drain or heavy leachate from rockwool slabs running off into the environment. Plants grown in soil tend not to produce as much odor while growing as a hydro plant, due to the plants transpiring less. This, along with the fact that there are usually no irrigation timers to malfunction, have given many a grower a better night's sleep. How many days can an aeroponic unit keep your plants alive without power?

## WHAT IS SOILLESS MIX?

When we talk about soilless mix, we are generally talking about a peat moss-based substrate with additives that vary between brands and mixes. The most popular North American brands like Sunshine Mix and Pro Mix usually have sphagnum peat moss as the body of the mix, chosen for its water holding abilities, relatively inexpensive cost and high "cation exchange capacity" – ability to hold nutrients for plant uptake.

Lime is added in two forms to raise the acidic pH of peat moss and supply a base of calcium and magnesium to the crop. Hydrated lime is usually added in very small amounts to provide long-term pH adjustment.

Perlite is added in varying sizes and quantities to make for a more porous mix. Wetting agents are used to help break the surface tension of the water and wet the mix more evenly upon irrigation. A starter charge of fertilizer is usually added to an EC of about 1.5 to provide enough nutrition for the first few days of growth.

Some brands offer mixes meeting organic standards with the main difference being the use of an organic wetting agent and a small amount of natural source fertilizer.

When buying soilless mix, it is best to stick to the big names like Premier, Berger and SunGro. There are a lot of garbage soilless mixes out there, and many vastly different qualities of peat on the market. Even among the big companies, the grade of mix that is sold in many retail garden center chains is not the same as what is sold to commercial greenhouses or nurseries.

The best quality peat moss comes in long strands, which translate into more air spaces in the root zone. One of the best byproducts of making top quality peat mixes is an abundance of peat dust and very small peat fibers. This is what soilless mixes sold in garden centers are made with. The problem with this is that the small fibers compact too heavily and do not let as much air into the root zone as a higher grade would.

The best bet is to buy the large 3.8 cubic feet compressed bales of soilless mix, as these are the ones that are usually top grade.

## 3 TYPES OF WATER

Regardless of whether you use drippers, water wands, or just a plain old watering can, understanding the properties of peat moss will make you a better gardener.

There are essentially three types of water that can be present in a soilless medium: unavailable water, available water, and free water. Unavailable water is water that is present in the medium, yet unavailable to the plant's roots because it is held tightly to the peat. Peat needs to be hydrated to a certain point before it has excess water that can be taken up by the plant.

Available water is that which is present over and above this amount.

Free water is water that fills the pore spaces in the soil, displacing oxygen. This type of water is something that should be minimized wherever possible, as this is what can damage roots, causing Pythium and other root diseases.

Photos:
Soil/seedlings:
Pete Brady
Garden products:
Andrew Young

# sensi

BY DMT

The only time that free water has a place in pot growing is during the first few days of seed germination until radical (root tip) emergence, or when some amount of leaching is required. Even during leaching this should be minimized by using the methods described below.

## WATERING PEAT

Peat has a waxy coating that actually repels water; this effect is magnified when the peat is dry, as a tighter waxy layer is formed. This is why if you water a very dry pot most of the water will run down the sides and out the drainage holes without wetting the medium.

Both to limit the amount of free water in the medium and to work around peat's water-repelling abilities, staggered watering techniques work well. By applying the desired amount of water spread out over two or three waterings at 20-30 minute intervals you accomplish two things: First of all, the surface tension of the peat is lessened after the first application so that the peat can more readily absorb the applied water. Second, free water in pore spaces has time to disperse through the medium or out drain holes, which minimizes the displacement of oxygen in the root zone.

When purchasing drippers, the best types are ones that are pressure regulated, which means that a specific pressure of water must be on the emitter before it will open. This eliminates the common occurrence of plants closest to the reservoir receiving more water than the ones at the end of the line. Netafim is a common trade name for this type of dripper although several other companies manufacture them.

Staggered irrigation cycles are especially important when using drippers, as all of the water has to disperse from a single source point.

Capillary matting, trough systems, flood benches and wick systems all allow pots to be watered from the bottom. This offers advantages in convenience, minimized soil compaction and less disease in some cases as the foliage remains dry. The main problem is that the fertilizer salts accumulate in the top inch or so of the medium rather than being flushed out the bottom of the pot as in conventional practices. Soil tests taken at the bottom, middle, and top portion of the pot can be substantially different. To prevent a build-up of salts, water the containers from the top about every ten days. The dried nutrients dissolve and are flushed from the top of the container and drain out.

Capillary matting can also be used very successfully as a complement to top-fed systems, especially in "sea of green" applications. Using capillary matting in conjunction with drip or wand watering works to ensure the entire crop is evenly watered, as pots that are a little drier than the rest can absorb the balance of water from the mat. Whenever possible use the porous black plastic mats designed to cover capillary mats as they cut down on algae growth as well as insect populations.

## WET/DRY WATERING

There are two limiting factors to the marijuana plant's growth rate which need to be kept in mind when watering, and unfortunately the two work against each other. Adequate water is required for plant growth, yet having a wet medium limits the amount of oxygen in the root zone, which

*Your plants should be just slightly pot-bound by the time they are going into heavy flower – this seems to give a slightly fuller bloom. Anything more than this is a waste of soil.*

reduces growth rates. A good wet/dry watering cycle maximizes both of these factors.

By irrigating until there's a slight leach (or a heavier one if you have bad water), and then letting the soil dry out before watering again, you are balancing both of these limiting factors. This is not to say that the wet/dry cycle cannot be deviated from – slight water stress is a great tool in controlling internodal stretch during the vegetative or early flowering phase.

There are also times in certain grow situations when keeping the soil fairly wet to encourage internodal stretch may be advantageous. For example, if growing a tight-noded Indica Screen of Green style you may want to encourage longer internodes during the late vegetative/early flowering phase in order to fill the screen more efficiently.

Either keeping the soil too wet, or more commonly, allowing the soil to become too dry between waterings both have negative effects on yield during bud formation.

## SYNTHETIC VS ORGANIC

One of the best things about growing in soil is how conductive it is to using organic source nutrients. Certainly it is possible to use organics in a hydro setup, but many who have tried simply find it to be too much trouble.

The organic versus synthetic fertilizer argument has been covered many times before, both in marijuana as well as normal gardening and farming circles. Proponents for synthetics often use the argument that a plant cannot tell organically derived nutrients from their synthetically derived rivals. Although this is true, what they don't mention is that with both organic and synthetic fertilizers, you are exposing your plants to much more than what is shown on the label.

The main downfall of fertilizers is in the heavy metal contaminants like cadmium, lead, arsenic and zinc, which are often present. Cadmium and arsenic for instance have been shown to cause cancer as well as other problems.

Synthetic proponents often argue that organic fertilizers come from unknown sources which may contain many of the aforementioned contaminants. Levels of these contaminants vary considerably between fertilizer brands rather than strictly along organic and synthetic lines. Proof is in the pudding, however, and

recently the state of Washington has required all fertilizer companies to show the pudding. Take a look at their web site: http://agr.wa.gov/pestfert/fertilizers/default.htm and do the detective work yourselves. Most of the popular brands used by marijuana growers are listed along with their respective heavy metal contents.

It would seem that marijuana would be more likely than many plants to cause ill effects on humans from heavy metal contaminants, both because of its efficiency at accumulating them and because the product is often inhaled directly into the lungs. Cannabis is well known for its above-ordinary ability to accumulate heavy metals as it has been used on many occasions to clean up contaminated soils. To my knowledge, no studies have been done on where these heavy metals accumulate within the plant.

## FERTILIZER APPLICATION

Organic fertilizers can be applied to soilless mixes in liquid form, by pre-mixed powdered slow release nutrients, or a combination of both. Most of the liquid mixes that have been around for a while, such as Earth Juice, Pure Blend and others, are safe bets.

Fish emulsion formulas work well and are inexpensive, although they likely should not be used exclusively. Out of the organic source fertilizers commonly used to grow weed, this is probably the highest in heavy metal contaminants, as fish are well known to accumulate mercury and other metals. Some growers also claim that it affects the taste of bud when used during flowering.

All-Mix or "Super Soil" recipes abound among growers and on the Internet. Some are good and some are not, so do your research and you should be able to find a good one. Some hydro shops also sell pre-mixed products for this purpose.

There are many synthetic fertilizers available on the market, many of which are tailored to meet the needs of marijuana (although this is likely not mentioned on the label!). Stick to the popular brands or ones that are popular in your area.

## LARGE-SCALE FERTILIZATION

Large-scale growers who wish to be able to tailor nutritional needs more carefully to the strain and growth stage may wish to try using bulk greenhouse grade fertilizers.

These come in bags such as 10-52-10, 14-0-14, 20-10-20, or can be custom-made from raw salts such as calcium nitrate and monopotassium phosphate. These items can be obtained inexpensively and without hassle from your local agricultural suppliers, but they usually come in 15 kg bags minimum.

Fertilizers similar to these are usually available in your local nursery as "all purpose" fertilizers. All of these can be used successfully to grow marijuana, but must be used in conjunction with calcium and magnesium based fertilizer like 14-0-14, which is usually only available from agricultural suppliers in the large bags.

These types of "all purpose" labeled fertilizers common in garden centers contain little if any calcium or magnesium. This is not an issue for people growing houseplants, as the lime in the soil provides sufficient quantities. However, for fast growing plants like marijuana, calcium and magnesium based formulas must be used in conjunction.

The cal/mag formula can be either mixed with the other fertilizer in appropriate amounts in the reservoir, or applied on an alternating watering schedule. For instance, during the veg stage, water once a day with 20-10-20, then the next day with 14-0-14. Or flower one time with 10-52-10, and one time with 14-0-14. The frequency of the 14-0-14 applications can be staggered depending on crop stage or desired fertilizer ratio. It does not have to be on a 1:1 ratio.

Newer growers would do best to stick to tried and proven pre-made formulas (General Hydroponics, etc.) from your local hydroponics store if growing with synthetic nutrients. Many of the organic formulas described above are also very beginner friendly.

## MIX MODIFICATION

When soilless mix suppliers prepare their mixes they are not gearing them to any specific plant and certainly not the needs of cannabis. The pH and porosity is designed to provide adequate conditions for a wide variety of plants. Some plants like a low pH, some like a high pH, some like a wetter root environment, and some like it drier. Most soilless mix takes a middle ground on these issues.

All of these factors can be easily adjusted to better suit the needs of cannabis. Extra dolomite lime can

*Whether indoors or outdoors, soil is always the most environmentally-friendly and forgiving medium to grow in.*

be added to raise the pH from the usual pre-adjusted 5.8 into a more acceptable 6-7 range (6.5-7 seems to work best for organic nutrients). Additional perlite will increase the drainage and amount of air space in the medium more to Mary Jane's liking.

### TESTING SOIL

Cannabis grows best in soil at a pH between 6 and 7 and an EC of 1 to a low 2, depending on the strain. The absolute lowest you should ever let your pH drop to is 5.8, as below that toxicity problems are likely and accumulation of heavy metals in the bud become more probable.

Both pH and EC in the soil can change drastically depending on the water source, nutrients and other factors. High phosphorous synthetic "Bloom Boosters" can easily make soil pH drop drastically especially if using cheaper grade fertilizers, which do not have "pH stabilizers" included in the formula.

Many veg formulas with large amounts of urea or ammonium nitrate in the formula can also cause rapid pH decline. Water or fertilizers containing high amounts of calcium or potassium generally increase the pH of the medium over time. Most of the higher end hydroponics fertilizers on the market are well balanced and adjusted to help avoid wide pH swings.

Always remember that in soil the pH or EC or the irrigation water is not as important as the actual pH or EC of the medium. Medium conditions can be measured by catching leachate for testing, using high quality soil probe meters, or mixing a soil sample with distilled water before measuring. Leachate tests can be considered fairly accurate at determining the actual pH and EC, provided it is not a heavy leach, which may alter the readings. Around 5% leaching is about right for this.

Soil probe meters are available for both pH and EC, although they are a fair bit more expensive than a normal meter. These probes will give the most accurate reading, but the cheap versions are no good (anything under $100 is suspect).

As distilled water has an EC of 0 and almost no buffer capacity, mixing a soil sample at a water to soil ratio of 2:1 will provide an accurate yet diluted EC measurement and a very accurate pH indication. Let the mixture sit for twenty minutes then take a measurement using standard meters or test strips. The pH can be taken as is, but the EC must be multiplied by 2.4 to take into account the pore space and dilution factor.

### TEMPERATURE CONTROL

Soil temperature is just as important in soil growing as solution temperature is in hydro growing. Many garden supply stores sell soil or compost probe thermometers, which are inexpensive ($30-$40) and well suited to this application. They can be left permanently imbedded in the medium or used as needed. Keep in mind that they take a few minutes to display the correct temperature after being inserted.

Proper soil temperature should be constantly maintained at around 70°F (21°C). Allowing soil to sit cold even for the first couple of hours before the lights warm up the pots can cost you in yield, as it makes it harder for the roots to uptake phosphorous. Usually just keeping the plants off of cold concrete floors or insulating them with foam panels is sufficient to keep temperatures warm enough.

In colder rooms with no night cycle heating, it may be necessary to use a bottom heat source. An easy method to create bottom heat is to fill a shallow tray with moist sand and run heat cables through the sand. The pots can be placed directly onto the sand or on a plastic covering.

### BAD BUGS

One downfall to growing in soilless mix is that bugs and their eggs may enter your grow room with the soil. The soil does, after all, originate in a peat bog somewhere. Although some people claim that spider mites were brought in with their soil, this is the exception rather than the rule. Fungus gnats, however, do occur with some regularity.

Most batches of soil will be free of these pests, and even when they are present they are quite easy to deal with. Mixing diatomaceous earth into the top half-inch of soil in the pot gives good control. Drenching the medium with organic pesticides such as *bacillus thuringiensis* variety *israelensis* (common trade names are Gnatrol or Vectobac), neem oil or SM90 also gives excellent control.

Fungus gnats can spread diseases and viruses and certainly are not a tasty addition to a nice cone of kind bud. Keep an extra careful eye out for these critters when rooting cuttings directly in soil as the larvae can burrow into the wounded base and cause serious damage or death to the cutting.

## POT SIZE AND COLOR

Pot size will be determined both by your strain and growing cycle. Sea of green growers usually use anything between one and three gallon pots, while bush growers use pots up to about 10 gallons. Ideally, your plant should be just slightly pot-bound by the time they are going into heavy bloom. It is not clear why this is, but it seems to give a slightly fuller bloom. Anything more than this is a waste of soil.

Transplanting should be done when a firm root ball has been formed and smaller feeder roots have filled out to the sides of the pot. Using several container sizes before the final one provides a much denser root ball, which in turn provides better buds. Roots tend to shoot straight out for the sides of the container and then circle it, leaving much of the area in between relatively void of roots. For example, by transplanting a clone in a 4" pot to a 6" one gallon pot before its final 8" three gallon pot, you will achieve a much denser root ball than if the 4" pot was transplanted directly into the 8" container.

Square pots are always better than round as they hold more soil in the same square area of space. Many growers debate whether white pots or black pots are better, but for the average indoor grower color likely makes little difference.

Black ones will get hotter (not always a good thing) and block light from reaching the roots, while white ones stay cooler yet may let some light into the outer perimeter of the root zone.

Clone growers may prefer to grow in large bed-type systems where many plants are planted in the same tub. This allows more room for root growth and avoids the circling problem associated with pots.

## ENVIRONMENTAL FACTORS

Peat moss is not a renewable resource. As I like to think that a good portion of marijuana growers are environmentally conscious, it is in everyone's best interest for growers to take a good look at some alternatives.

Coconut coir is a good substitute that many growers have been experimenting with or switching to altogether. A byproduct of the coconut industry, coir is a sound alternative to peat. The main problem experienced by growers so far is that coir can have higher levels of sodium than ideal when harvested from coastal areas. Any good grade of horticultural coir should be low sodium.

Some non-cannabis growers have also experienced problems in rooting cuttings of some types of plants, presumably from some type of hormone present.

Yet aside from its environmental aspects, coir has a much better starting pH than peat (6-7) and is more porous. As with anything new in growing, don't jump right in until you have done some trials to make sure it will work properly with your given conditions and techniques.

Growing pot organically in soil is one of the finest examples of the benefits of keeping things simple. The world would be a better place if more people rolled up their sleeves and dug around in the dirt. 🍃

• Washington State fertilizer database: www.cannabisculture.com/news/fert

story and photos
BY TINA LEE

A simple way to keep well-watered while

# The wick

**Y**our buds are just starting to ripen, and you are looking forward to a great harvest. Then your friend gives you a call, inviting you to a fabulous once-in-a-life-time long weekend gathering. You look at your garden. Your plants need to be watered once a day. You can't go!

. . . Or can you? Sooner or later everyone who grows marijuana will need to spend a few days away from their plants. It can be done! How? With the Wick Method, of course.

## why wicks?

Some growers like to use the wick method because it means a minimal amount of effort when it comes to watering. The wick method allows a grower the ability to minimize the number of times that she is required to water her plants. Almost any growing container can be adapted to the wick method. All that is required are a few nylon wicks about an inch to an inch and a half wide, suitably cut to length; a container to hold water (a reservoir) which is larger than the container that your plant is growing in; and a device, such as an inverted container, to stand your plant container on.

The advantage of the wick method is simple. It means more days on the beach for me. A supply of water for up to six days or more can be made to slowly release itself to any given plant. The amount of water released to each plant can be regulated by how many wicks a grower uses. Using the minimum of one or two wicks per container delivers the minimal number of water a plant needs. Therefore the reservoir the plant is using will deliver the maximum amount of watering days. The number of days a reservoir will supply minimal watering requirements is regulated by the size of the reservoir used, and the amount of water in the reservoir.

If you are going to be away from your plants a few days, you can make a wick system in advance and test it before you leave to see how many days it will work. This will help you to determine what size reservoir you will require, and how much water the reservoir will need.

## grow container

If a grower needs to leave his or her plants for a few days, they need only to adapt their growing situation to this simple method, and their watering worries are over.

In order to adapt container plants to the wick method, the nylon wicks may be placed in the containers when the soil and plants are first put in the container. However, the wicks may also be easily added later, using a thin stick such as a ruler to push the wicks down the inside edge of the container, through the soil.

The wicks should be placed along the inside of the containers, running from about an inch from the top of the soil, along the inside of the container, and down and out the drain holes at the bottom of the container. The wick should hang down and out of the bottom of the container, and be as long as is required to reach the bottom of whatever reservoir it will be placed into.

## reservoir readiness

Once the wicks have been placed in a grow container, the growing container must be placed in a

# method.

your potted plants
you're away.

reservoir of water. The container in which your plant is planted does not touch the bottom of the reservoir in which it is placed. The container must be on a stand or be wired in place, so its bottom is an inch above the top of the water line in the reservoir.

The bottom of the plant container never touches the water in the reservoir. That is what the wicks are for. Only the wicks, protruding from the bottom of the growing container, will be placed in the water. The wicks should extend out of the bottom of the grow container and go all the way to the bottom of the reservoir.

## ready to go

The budding plant pictured here has been planted in a container which holds about two gallons of soil mixture, with a layer of lava rock on top. The wick method allowed this budding plant to bloom for a few more days. I picked her when I returned from a weekend at the beach, just as she reached her maximum potential.

The other two photos show the plant and its "wicked" container inside the reservoir. There is a small cup inside which sits inverted, and is just high enough so that it keeps the bottom of the plant container an inch above the water level in the reservoir. When the water in the reservoir is filled to the water line

*At right: These reservoirs have a hose attached to make watering even easier.*

there is about one gallon in the reservoir.

This particular plant uses three wicks. This is ample water, which will capillary to the plant, to last for about three days. If I wanted to lengthen the time of automatic watering, I could raise the plant container, add more water to the container and extend the wicks. This reservoir could hold enough water to supply a plant for a week or longer.

I love indoor growing. However, it can be very demanding, needing almost daily attention. The wick method gives me the ability to enjoy a few days away from the buds every now and then.

So the next time you need a few days away from your plants, or if you just want to make watering easier, try this unique wick method. It is simple yet effective, and can bring you peace of mind.

▲ *This plant is ready to go on a stand in a reservoir. These wicks will feed it water for days.*

## STANDARD HYDROPONICS – STANDARD PROBLEMS

In the standard "ebb and flow" hydroponic system, there is an upper grow bed which is used to hold the plants. This is a box which holds a certain number of plants in containers. Different mediums may be used in the containers, including lava rock, rockwool, perlite, vermiculite, coconut fiber, and even Styrofoam pellets. The best medium is one which retains a small amount of water for a long period of time. Porous materials such as lava rock are excellent.

There is also a lower reservoir which holds the nutrient-rich water. At regular intervals nutrient water is pumped upward into the grow bed using a timer and aquarium pump.

The main problems most novice hydroponic growers have is maintaining the proper nutrient levels in the water reservoir, keeping the ratio of nutrients correct, and having a constantly balanced pH. If these three things are not precisely calibrated, the hydroponic garden will not thrive as it should.

Specialized concentrated nutrients are used in the reservoir, because standard organic nutrients will interact with other organisms in the water and begin to decompose. This can quickly make your

# ORG HYDRO
## A SIMPLE WAY TO USE ORGANIC NUTRIE

*top, left to right:*
*Budding plants at various stages all share space, light and water.*
*Looking down on an organic hydro garden.*
*Roots emerging from their container.*
ALL STORY PHOTOGRAPHS BY HANS

reservoir water into a rotting, toxic soup. The concentrated, "clean" synthesized chemicals do not interact with organisms in the water environment, and therefore the reservoir water stays "clean," more or less. However, these "clean" chemicals are less forgiving than organic fertilizers, so that over-fertilizing will immediately burn your plants, before you have a chance to notice they are beginning to burn.

As the nutrient water saturates the roots, the plant and the grow medium both retain some of the nutrients. The water that returns to the reservoir therefore has less parts per million (ppm) than the water had before it was cycled through the upper grow bed. This means that the nutrient level and pH of your reservoir is in constant flux, and requires careful, persistent monitoring for a successful grow.

## ORGANICS IS EASIER

Growing marijuana hydroponically with totally organic nutrients is actually easier than growing hydroponically with man-made chemicals. Because we do not place any chemicals in the lower water reservoir, we remove all of the problems

*bottom, left to right:*
*Fresh cuttings preparing to root.*
*An organic garden of hydro delights!*
*Pumping up a plant.*

ANIC
PONICSS

WITH A HYDROPONIC SYSTEM.   BY HANS

associated with maintaining the proper ppm of the water. Eliminating the need to balance different nutrient levels with ppm and pH in the water reservoir eliminates most of the problems associated with hydroponic growing.

With my organic hydroponic technique, the reservoir water is applied to the upper grow bed in the same way as usual, with a timer and an aquarium pump. But the reservoir contains only pure water

the lower water reservoir.

However, if the grow containers are sitting in water, then when the water drains back down into the lower reservoir, the water will contain nutrients leached from the medium. The reservoir would soon become polluted. The secret to avoiding this, the secret to successful organic hydroponics, is in the construction and placement of the grow containers in the upper grow bed.

the water when it's pumped to the upper grow bed.

Also, the bottom of the plant container should not touch the bottom of the grow container. This is best accomplished by placing each plant container on a small stand, such as a small plastic cup or even a few extra pieces of lava rock. If the bottom of the grow container is resting on the bottom of the grow bed, it will cause a backward capillary action of the water

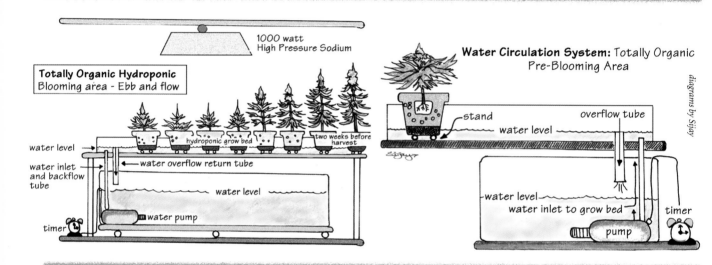

Totally Organic Hydroponic Blooming area - Ebb and flow

1000 watt High Pressure Sodium

hydroponic grow bed

two weeks before harvest

water level

water inlet and backflow tube

water overflow return tube

water level

water pump

timer

Water Circulation System: Totally Organic Pre-Blooming Area

stand

water level

overflow tube

water level

water inlet to grow bed

pump

timer

diagrams by Sijay

– the organic nutrients are applied to the top of the medium, the same as you would water nutrients into a container of soil. Nutrients are mixed with a cup of water and applied by pouring the cup evenly over the growing medium.

The nutrients are applied when the upper grow area is dry and there is no water in the grow bed. That way, if a small amount of nutrient water should leak out of the bottom of the grow container and onto the bottom of the grow bed, it can easily be sponged up before the next watering cycle begins. This will keep the nutrients in the containers of medium and not in

## CONTAINER PLACEMENT

Any size of grow container may be used. I take standard 6" or 10" diameter plastic grow containers and place lava rocks or pea gravel in the bottom, filling them up about a quarter of the way. The remainder of the container may be filled with the medium of your choice: rockwool, coconut fiber, or my favorite, a combination of pre-wetted perlite/vermiculite/lava rock.

The grow container is placed in the upper grow bed in such a manner that only the bottom two inches or so, (the part which has lava rock in it) will come into contact with

in the plant container, drawing nutrients out of the medium and into your reservoir.

The water level height in the grow bed should be regulated by the use of an overflow tube. This allows the maximum water level in the grow bed to be set at the necessary height.

As the grow bed is flooded with pure clean water, the lava rock in the containers will become saturated and begin to draw up the clean water by capillary action, into the nutrient laden, almost dry medium. After the clean water has circulated in the grow bed for half an hour, the water is returned to the lower reservoir, as clean as before it began the cycle. The water is still clean because when water is capillaried upward, it does not saturate the medium which holds the nutrients, so no water will flow back downwards.

Once the water stops circulating, the lava rock in the bottoms of the plant containers will still be saturated, and will continue to capillary water in an upward direction. Once the lower lava rocks dry out, the upper medium will begin to dry more. (Be sure the bottom of the plant container does not sit directly on the bottom of the grow bed!)

Once the upper medium has reached a certain level of dryness, it is time to water again. The water is cycled again to the upper grow bed and the process begins over. The container size, medium and other environmental conditions will dictate the timing between watering.

There is an easy way to determine your watering schedule. Simply do not water your plants, and watch for the first one to wilt. If your plants can go six hours before the first one wilts, then set the timer and pump to water every four hours. Be sure to check and update your watering schedule as the plants develop.

## NUTRIENT APPLICATION

Nutrients are added to the grow medium as needed, every few days. Any organic powder may be mixed with a cup or so of water, and poured directly on top of the growing medium. If the instructions for a certain organic nutrient say "use 2 tsps in a gallon of water every two weeks," then try taking 1/2 tsp and add it to a gallon of water.

Every 3 to 4 days pour about a cup of this into the top of the medium in each container. If some drains out the bottom, then next time use less than a cup, until you find the perfect amount to use.

The amount of nutrient each plant requires can be determined by increasing the dose to a single plant. Watch for the increase in the deepening of the green color of the leaves. Darker green is an indication of maximum feeding. If the plant gets too dark and the tips of the leaves begin to turn under and burn at the tips, you will know that the plant in question has received the maximum schedule of nutrients. Give all your other plants a little less nutrients than the one that was beginning to burn.

It doesn't take long to figure out how much nutrients should be applied to each plant. The great thing about this method is that you are unlikely to burn any of your plants or underfeed them.

When you want to flush your plants before they finish, you simply reduce or eliminate the amount of organic nutrients to any given plant during the last two to three weeks before harvest. When the plants are deprived of their nutrients, they begin to naturally draw on the stored reserves which are in the larger food leaves. As the stored nutrients are used by the plants during the last weeks, the leaves which give up their stored nutrients begin to change color. This is what happens in nature during Fall.

This means that the smoker will not be smoking the unpleasant and unhealthy man-made chemicals which are present in plants grown in the standard hydroponic fashion. A plant which has had the opportunity to be grown in a totally organic environment, and allowed

to use up its stored nutrients before harvest will not only taste and smoke much better, but will be healthier and better for you.

This process may be used with seedlings, pre-blooming plants and blooming plants, as well as mother plants, which may be grown for longer periods of time. This totally organic hydroponic technique may be used with either the "ebb and flow" or the "nutrient film/flow" techniques. It is medium friendly, and can use standard nursery items as well as specialized mediums which one would find in grow stores.

So throw away those man-made chemicals, ditch the expensive ppm and pH meters, and forget about keeping proper chemical balances in your reservoir! Grow organic hydroponics, for some sweet and easy bud! 🌑

• This process is one of many detailed in Hans' book, *Growing Marijuana Hydroponically*, Ronin Press.

# Cannabis Culture:

*Indian Black*

*Nepalese Temple Ball*

*Bubblehash*

**Story and Photos
by Pete Brady**

THE MODERN HISTORY OF Afghanistan is permeated with cannabis and conflict. This country of 25 million people is impoverished, demonized and flattened by war, lacking permanent water supplies and surrounded by hostile neighbors who shut out its refugees. Like the other nations in this region—India, Iran, Pakistan, Kashmir and Nepal—Afghanistan has marijuana traditions that span centuries and embody the highest arts of cannabis production, processing and consumption.

The British ran Afghanistan for decades before they were kicked out in 1919. The country was relatively stable during the reign of King Mohammed Zahir Shah, a pro-cannabis monarch who governed Afghanistan from 1933 until he was overthrown by a jealous relative in 1973.

According to reports from U.S. spy agencies and Afghan sources in Holland, the king offered armed protection and horticultural advice to marijuana growers, encouraging them to increase their yield with modern fertilization techniques. The ruler's top aides were allegedly involved in

overt hashish smuggling. DEA officials even allege that the king's private jet was used to smuggle tons of hashish to Italy and other European countries.

During the 1960s and early 1970s, Afghani hash was considered the best available. Cultivation of squat, rugged, fat-leafed Indica plants, which cannabists now call "Hindu Kush," "Afghani," and "Hashplant" became prevalent during this era; some ethnobotanists say Afghanistan's earlier cannabis farmers mostly grew Sativa varieties.

Western hippies collected Afghan marijuana seeds and spread them across the world in the 1970s, most notably to Northern California, where the seeds became genetic precursors for many of today's most popular cannabis cultivars, including "Skunk."

Afghani hash was known for its sticky, resiny, unadulterated color and texture, its sweet tangy taste and its narcotic, dream-inducing high. Before U.S. anti-drug pressure changed Afghanistan's cannabis policies in 1974, super-potent connoisseur hashish was available at

# Afghanistan

*Left:*
*Poppy flower*

*Afghani Chunk*

*Ice-o-lator hash (background: wet/unpressed)*

teahouses inside Afghanistan, and as exported fingers, sticks, hooves, half moons, slabs and bricks that had a wide array of colors, tastes and cannabinoid profiles.

After King Zahir Shah was deposed, the U.S. began sabotaging the Afghan cannabis industry, beginning a series of intermittent drug wars in Afghanistan. The U.S. paid Afghan governments millions of dollars to eradicate cannabis crops and hash producers beginning in the mid-1970s. The elimination of ganja farming and hashish production cost lives and money, spurred production of opium poppies, plunged a poor country further into poverty, and also resulted in numerous human rights violations.

By the time the country was invaded and occupied by the Soviet Union in 1979, the Afghan cannabis industry was a mere shadow of what it had been.

In 1989, the Russians fled Afghanistan, leaving their soldiers' blood and thousands of live land mines behind. *Mujahadin* factions fought amongst themselves for control of the war-ravaged country; the ultra-fundamentalist

Taliban won the power struggle and established a theocratic government in Afghanistan in 1996.

The perils, contradictions and ironies of the drug war are starkly outlined by U.S. policy failures in Afghanistan. The U.S. created, trained and outfitted the Taliban, Osama bin Laden, and Al-Qaida from 1978 to 1990, when Cold War realities made it important for the U.S. to help Afghanis throw Russian occupiers out of Afghanistan. There is evidence that the U.S. still had ties to bin Laden and the Taliban, even as recently as September 2001, and beyond. The U.S. gave $40 million in drug war money to the Taliban, supposedly for eradication of poppy fields, in summer 2001.

The subsequent U.S. invasion of Afghanistan, the overthrow of the Taliban regime, the U.S.-UN occupation of Afghanistan, and the ongoing strife there (in which United Nations peacekeepers, aid workers and others are killed weekly) has only exacerbated an already dangerous situation.

And through it all, Afghani farmers have still been able to plant massive crops of opium poppies and marijuana. The traditional

trade routes for drugs, via Pakistan and Central Asia republics, have not been disrupted; coffeeshops in Holland are again advertising fresh Afghani hashish for sale.

Today's Afghani hash is considered a mid-grade product, slightly inferior to primo traditional hashish from Morocco, Nepal, India and Europe. It is only about 40% as potent as the newest types of hashish, such as water hash. 🌀

Afghani hash

# Aeroponic Supersonic

*Aeroponics allows for maximum oxygen and nutrient absorption, which creates dense, compact and abundant buds that will amaze you.*

## by Ashera Jones

I first heard of aeroponics through a friend, who told me of military experiments in growing enormous tomato plants indoors by means of aeroponics and cooled lights. When I learned that the roots just hang in the air and are misted by atomized water and nutrient particles, I thought "how unnatural."

Then I saw a system set up: it was so clean and efficient! I saw that this was a very effective nutrient and oxygen delivery system; one which, by nature, is quick to respond to whatever nutrients you introduce into the system, making both feeding and curing more time-efficient and accurate.

But is aeroponics really unnatural? Perhaps not. My aeroponic system reminds me of my visit to floating coral islands in the Bahamas, where the coral provided a natural aeroponics system, and everything grew to an enormous size. The oranges were the size of grapefruits, and the hummingbirds were the size of robins.

The layout is important. Other than some basic plumbing supplies, and low cost but reliable

*Aeroponics can get big buds out of small plants.*

sump pump (otherwise known as a jet pump), the lights, environment and electrical are all the same as a regular grow op. One of the advantages with this system is that there is little to dispose of later—this brings less police heat and so less stress for you.

You can grow aeroponically in tubes, buckets, or anything that is sealed and opaque. There are kits and contraptions you can buy, or with a bit of effort you can compose your own. Be prepared for some watery catastrophes

and have a wet and dry shop vacuum on hand. Make sure your floor is protected.

## The 4 Elements of Aeroponic Gardening

### AIR

In an aeroponic system, nutrients and water are sprayed onto the roots in an atomized or mist form by a high-pressure pump. This creates quickly moving water

## Hydrogen Peroxide (H₂O₂) or "Oxygen Water"

Hydrogen peroxide created by humans is inferior to that created by nature, but it costs less. Human-made peroxide comes in a variety of percentages, and you want to get the 35% variety, as this will ensure that there have been no "stabilizers" added, as is generally done to the 3% variety that you can buy in a pharmacy.

Hydrogen peroxide will most definitely reduce any possibility of bacterial and fungal contamination. Farmers in the U.S. use peroxide for everything from disinfection of the dairy barn and hog pens to increasing crop yields by 20 to 30%. It is also commonly added to the drinking water of animals to reduce the need for antibiotics.

In the aeroponic system, $H_2O_2$ replicates nature's own antibacterial mechanism and prevents water from growing putrefactive bacteria which can cause the dreaded bacterial wilt, root rot and countless other diseases.

Thirty-five percent hydrogen peroxide can and should be used in hydroponic drip and soil systems at the rate of one teaspoon per gallon. You would usually add peroxide to your system and let that run through for 1/2 hour before adding nutrients. This will give the peroxide time to kill off bacteria with its extra oxygen molecules and become stabilized before you add the dissolved mineral salts (stock fertilizer). $H_2O_2$ will also forcibly cause the out-gassing of chlorine and fluoride from municipal water.

I have used 35% hydrogen peroxide diluted to a 1% solution on a plant in soil that had an infestation of fungus gnats, root rot and unknown other problems. The plant thrived while gnats and other organisms did not. But beware, $H_2O_2$ is powerful and experiments can be dangerous to your whole crop.

Be sure not to get any in your eyes. The burning and whitening sensation felt when it touches the skin can be irritating but is not damaging. Wear latex. You should dilute 1 part peroxide to 11 parts distilled water to create a 3% solution for less worrisome handling.

### Applying H₂O₂ to your System

You can kill bacteria that might be living in your nutrient tank by wiping out the empty container with a 3% peroxide solution. You can safely create a 1% peroxide solution to feed your plants for a few cycles, but fill less than a quarter of your nutrient tank with this solution, as you will want to dilute it to about 0.25% for longer use.

If there are parasitic invaders, you will know because there will be a profuse bubbling and frothing, which is the hydrogen peroxide oxidizing putrefactive organisms that are not oxygen compatible.

To make a 1% solution, add 35 parts water to 1 part 35% hydrogen peroxide. To achieve an 0.25% solution add 140 parts of water to 1 part 35% peroxide, or add 3 parts water to one part 1% peroxide.

*Healthy roots make happy plants and rapid growth.*

which is capable of delivering more oxygen because it is well agitated, like a waterfall. The presence of more oxygen also discourages bacterial and fungal growth.

The most effective root medium is the one which delivers the most oxygen to the roots. A dense soil may only deliver 30% oxygen to the roots, while a soilless mix will deliver up to 50%, and hydroponics will deliver around 80% oxygen. With aeroponics the sky is the limit, you literally receive 99% possible oxygen to the roots.

Aeroponics also allows nutrients to reach the roots directly, with no medium in-between to hamper nutrient uptake or foster the growth of bacterial organisms.

I have measured aeroponic plant growth

against soil, soilless mix and hydroponic drip. The aeroponic system doubles the growth rate of plants as compared to a soil system, and is about one-third faster than a hydroponic system. My experience is that it has been easy to grow fat, bushy, almost hardwood-stalked plants.

## EARTH

Just like all plants, plants in an aeroponic system fuel their growth with natural elements from the earth, which are easily obtained in liquid solutions. These store-bought nutrients are limited in content, and I will suggest other substances you can add to enhance the nutrient quality your plants are getting from aeroponic feeding.

The aeroponic garden prefers a lower nutrient solution of between 700-900 parts per million, and an acid pH of 5.5 to 5.8, which means that you will generally need to add a pH down. There are natural alternatives for the open minded that I will talk about later.

In the earth category, we also have the growth mediums, which although limited in an aeroponic system, are necessary. Personally, I use rockwool cubes and baskets, which vary in size.

The usual basket size used in hydroponic systems is 3.5 inches, but you can use 2-inch baskets for mature plants in an aeroponic

*This system has the stadium set-up, all plants growing towards a central array of Sodium and Metal Halide bulbs.*

system, because all the baskets and wool are doing is supporting the plants, keeping them stable and standing. However, I personally use the 3-inch size because I like to grow 3-foot bushes that become weighted with fat, juicy buds, and they require a little bit of support.

Nutrient uptake varies depending on the growth stage of your plants. I have found that the plants tend to uptake less nutrients when they are changing from vegetative to flowering, and at the same time, the type of nutrient that they require changes. It is essential to have a "parts per million pen" so that you can monitor the amount of food your babies are eating and prevent burning of the roots by a solution that is too acidic and full of mineral salts.

## FIRE

The element of fire in aeroponic gardening includes light and heat. Light brings photosynthesis and also ultraviolet "C" rays which discourage some types of bacteria. But excess

heat can bring fungus and bacteria, so a good balance is necessary. The layout of your room will determine light availability and the ideal situation is to have all plants receiving as much light as possible, without burning. This will also encourage the dense bushy-type plant that you want to grow.

I use a few strategically placed 400-watt lights, with the plants in a stepped, stadium structure around the lights. The placement of plants in what is known as a "stadium" ensures that I get the lights right in there, and give all the potential budding sites fair exposure, while discouraging phototropism, the search for light.

The ambient room temperature is no more than 80°F but preferably closer to 70°F (20°C). A cool room temperature usually means a nice, cool water temperature, which is also important.

## WATER

The element of water in aeroponics is the great transport system. I suggest you keep it cool and clean. Let water sit a day before adding it to the nutrient tank, and put in one teaspoon hydrogen

*These little clones will soon be sporting big buds in this tube system.*

peroxide per gallon of water, to make sure it is free from unfriendly organisms.

All that your plants receive comes to them through water. Without water there is no air.

To get your water to the roots, you will need a pump, a bunch of high-pressure spray heads and some hose.

People will try to sell you very expensive pumps, but you need spend no more than $200 to buy yourself a pump that puts out 60psi, which is appropriate for atomized nutrients. This kind of pump is available in the local plumbing department of a Home Depot or similar hardware store near you.

Your pump is meant to run intermittently. Do not bother using the pressure-tank system that can be purchased in conjunction with your pump. You will need an hour timer that will allow your pump to spray for 30 seconds to 2 minutes, and then rest for 4 or so minutes. This will prevent your pump from burning out and your system from flooding.

## ORGANIC AEROPONICS

Organic aeroponics is at least partially possible with little or no hassle. No indoor gardening can be completely organic. Aeroponically, the best plan is to introduce partial organics which provide micronutrients not available in standard inorganic, mineral-salt based, hydroponic mixes.

You can provide your plants with their requirements as organically as possible by making your own nutrient teas with plants that you know "fix" certain types of minerals into themselves. First dry the plants you want, then put them into water. Let them sit for a few days, then boil them, let them cool, put them through a fine particle filter, test the pH and parts per million so you know what percentages to add, and then introduce them to your aeroponic garden. The same can be done with "meals" like soybean meal and others.

I do not like to use blood and bone meals because I am not sure of the source. I would rather use plant sources of nutrients to help the persons who ingest my buds avoid contracting Mad Plant Disease!

How many people are using animal-based fertilizers whose origin they know nothing about? There is less restriction on the source of these "bone and blood meals" than there is in the pet food or agricultural industry, meaning that the indestructible "prion" particles responsible for "transmissible spongiform encephalophathy" (mad cow disease) are likely also present in bone and blood-meal plant fertilizers, and can find their way into plants fertilized with these products.

Liquid organic multivitamins for plants are useful and I would suggest using them in combination and in low concentrations. Be careful to avoid creating a giant toxic soup by adding nutrients all at once. Add one and then wait before you add the other. You will be amazed how readily your plants will gobble up the food you give them, and with aeroponics you can

*A plump bud grown the aeroponic way!*

expect near immediate results, good or bad.

## BAD RESULTS

If you have bad results after feeding, remove all nutrients and run a low peroxide solution through your system. Everything should be visibly back to normal within a day. If you see no positive results after a day, you may have a problem unrelated to feeding, or you may have added too much nutrient solution.

Too much nutrients and only the gods can save you. The gods, and maybe a little less lighting. Unplug a light or two and give your plants a chance to recover, as their energy will then be concentrated down in the roots, which will be trying to heal from chemical insult. If the leaves start yellowing, you have probably developed root rot. Shock can precipitate root rot and you must remember that plants have an immune system and do respond to stress, so try to avoid disasters.

*One of many beautiful buds you can harvest from your aeroponic system!*

The problem with introducing organics into an aeroponic system is that there is always some other organism that wants to cash in on the good life. To avoid such a problem, I run my organic fertilizers through the system first. Then about a day later, when I am low on nutrients and water, I drain the system and put my hydrogen peroxide solution in and let it run for 1/2 hour, sometimes 3 or more hours, depending on how much I have added. Then I add my stock nutrients, the ones you buy at the store. I let them run through for half a day, then I add my organics again. I let those run through for a day, the food and water get eaten, and then it is time to clear the system and add the hydrogen peroxide again. Hydrogen peroxide kills unwanted freeloaders.

## PH BALANCE

Aeroponically grown plants prefer a rather acidic solution of between 5.5 and 5.8 pH. With a 900 parts per million nutrient solution you will still need to add some pH down. A higher concentration of mineral salts generally makes the water more acidic and brings pH down and parts per million up.

What are in those pH up and pH down bottles anyways? Who cares! Nature's best solution is simple, cheap and has inherently less packaging. Lowering the pH can be done with apple cider vinegar but I like to use Kombucha fungus, as it creates a wonderful selection of living interacted nutrients that are amazing and affordable.

Basically, you grow a particular bacterium in a medium of black, green or herbal tea, and then add the altered medium to your solution. Kombucha is an oxygen producing bacteria that is compatible with both the human and vegetative world because it metabolizes nutrients in the tea itself, providing the plant with easy to assimilate nutrients. Kombucha also has beneficial health effects for the human organism.

Having an acidic, low-pH environment will reduce fungi like root rot. It is important to note that oxygen creates a high pH or neutral environment. So using hydrogen peroxide will bring your pH up, which is problematic because

"Kali weed" likes low pH in the aeroponic environment. By adding Kombucha, a living nutrient, you can lower pH while still providing oxygen and bringing micronutrients to your plants that they would not regularly get.

Another way to raise pH is by adding a small quantity of baking soda, and I mean small!

## CLEARING YOUR PLANTS

Before harvest, you will want to leach the unused mineral salts out of your plants. Leaching is easily done with aeroponics by changing the water daily for three to seven days. Note that you do have to change the water every day, as just running the same water through won't work!

Because mineral salts concentrate in the plants at a high pH, if you run a low pH through while leaching, your plants will release their mineral salts at a much faster rate. Kombucha with its low pH, is thus the ideal thing to add to your water during leaching. It facilitates the release of mineral salts, while also providing the plants with a continued, clean source of micronutrients.

## PHYTO ESTROGENS

Some people use birth control pills to raise the estrogen level in their plants. This is a personal choice, but it is not mine. These pills are synthetic and cause a lot of trouble for women and generally I have a certain disdain for them.

Menstrual blood, however, is a good source of estrogen and is as organic as the donor. Decreasing other fertilizer levels before introducing blood will reduce the possibility of the blood feeding unwanted organisms in your system. Ultimately, I can say that there is a part of me in every plant I have ever grown.

It is annoying that some, usually young, square men think this is disgusting. Get real. They prefer using bone and blood meal which they know nothing about, other than that it came from a dead animal who had a miserable life and ate less organically than most of the women you know.

Such backwards attitudes aside, menstrual blood is not plant-based and is arguably a secondary source

to phyto-estrogen. Blue Cohosh is an herb which contains plant estrogens, and I also use this in my flowering formulas. It can easily be introduced in the form of

tea, or you can even grow your Kombucha on it.

Try Aeroponics and have a "mist"-ifying experience!
SATIVA DIVA

*. . . can be used to feed eight tube ends each holding 20 pots!*

## Diseases Common To Aeroponic Systems

There are lots of other problems gardeners face, but these ones are particular to aeroponics.

### Fungus Gnats

Curing your aeroponic garden of diseases is very easy. Fungus gnats cannot get very far and are generally not a problem. They can only go a few inches into the rockwool before they meet the nutrient solution, which is no good for them. The standard powders, including diatomaceous earth, also work, so keep your eyes peeled for the evil, little, hopping flies and dust them with death at the first sign.

If you notice that you have more than five in your room, I would definitely run a 1/2% solution of peroxide and no nutrients through for one day to help the plants and kill any gnat larvae that may have made it into the medium. The larvae will also be filtered out by the fine particle spray filter before long.

### Root Rot

Root rot is particularly dangerous in aeroponic systems. It is a fungus that shows up as rust, a black spot and/or a powdery mildew. It might also come as a simple yellowing of the leaves and kill your babies in a short

period of time, depending on your response. A sure sign is a browning of the roots, not to be confused with the staining caused by certain nutrient solutions. Another sure sign is black spots on the roots, which accompanies the brown discolouration. Check your roots regularly.

There is always a margin of time before your plants will die, which can be increased depending on your knowledge. There are products which can be introduced through either the leaves or roots, both of which are effective.

One way to avoid root rot is to innoculate your plants' medium with a predator fungus. These are readily available at well-equipped grow stores.

*Long lush healthy roots*

### Bacterial Wilt

This is a weird one. Once you have it, pray and change everything. That sounds radical but so is the problem.

Bacterial wilt causes yellowing of leaves and flowers, and a portion of the stem directly below the yellowing will be entirely lifeless, brown and mushy, with a clear viscous goo oozing out of the plant's pores. Wilt is not common, and even in gardening books it is rarely mentioned, yet if environmental conditions are right it can affect your plants.

Not enough UV, and air which is too cool, may lead to bacterial wilt. Ultraviolet light is a natural part of the spectrum and kills bacteria. Water-cooled lights have a tendency to cut out a lot of UV. These cooled lights can be placed really close to plants, delivering lots of light, increasing both growth and resin production while countering the effects of photo and geotropism (the effects which lead to increased internodal spaces). Yet such lights negate the healing effects of both light heat and UV rays.

Hopefully wilt never happens to you. If it does, add a UV light or two. They are inexpensive and may give you a better chance at recreating what nature delivers.

*A small, inexpensive pump in a corner . . .*

# Cannabis Culture: MOROCCO

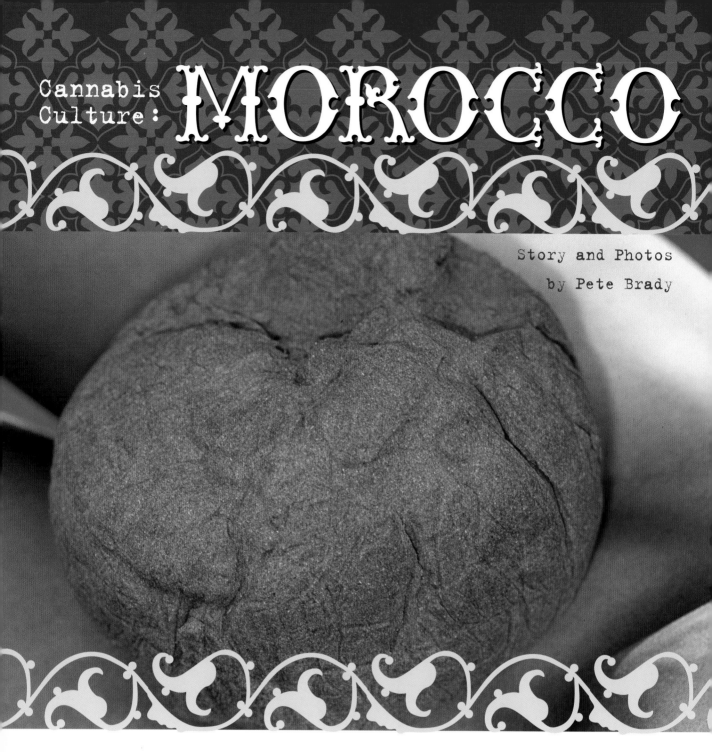

Story and Photos
by Pete Brady

In the marijuana scene, going to Morocco is kind of like making a pilgrimage to Mecca. Morocco somewhat resembles Mecca, with street signs encoded in ancient symbols and mosques on every corner, and *muezzin* singing prayers five times a day from those mosques.

The prayer songs echo in your head like hallucinatory scenes from the classic William Burroughs novel *Naked Lunch*, which was written while Burroughs was in Tangiers in a weeks-long trance induced by eating Moroccan hashish. Morocco is a marijuana Mecca because the North African country is the world's most consistent producer and exporter of cannabis resin, otherwise known as hashish.

Nobody knows for sure how much marijuana is grown in Morocco, how much land is devoted to cannabis cultivation, or how much Maroc hashish is exported each year, but estimates from the European Union, the United Nations and the CIA claim the country produces at least 30,000 pounds of hashish annually.

The scope of Morocco's cultivation can be seen in photos of huge clouds of male cannabis pollen that shroud Northern Africa and Southern Europe when the country's pot plants are in full bloom. An estimated 220,000 Moroccan acres are used for cultivating pot between early March and early September. Almost all of the acreage can be found in the rugged and remote Rif Mountains, located in the country's northeast quadrant.

The Rif is a spectacular geologic feature, stripped bare of its natural vegetation by centuries of overpopulation and irrational land use practice, so that the new undulations of stone, peaks, valleys and cliffs are seen with stark clarity.

The Rif is Africa's Jamaica—its fierce, resourceful, independent residents live for the most part in splendid isolation managing terraced farms, scant water supplies and ganja fields. The area's residents are called "Riffies," and they have their own form of superb indigenous music, known as "Rif rock" that serves as Morocco's equivalent of reggae.

Riffies are literally a breed apart. Their culture is tribal, based on the tradition of Berber warriors who have fought kings and drug warriors, forcing invaders down the mountains toward the Atlantic Ocean and Mediterranean Sea which border the country.

Besides its own music, the Rif has its own language and customs. Outsiders are not necessary or welcome. Riffies grow, process and transfer cannabis products to other Moroccans and to Europeans, who are largely responsible for getting Maroc resins across the water to Europe.

## From Cannabis to Kif

In Morocco, resin powder is produced using methods historians say have only been used in the country since the early 1960s. Before these methods were introduced, allegedly by Europeans and Northern Americans, Moroccans did not process cannabis into powder and hashish—they only smoked it as kif.

Morocco's hashmaking methodology involves placing marijuana plants on top of sieves, pieces of cloth or other materials that have very small pore sizes, covering the cannabis with heavy plastic, and then shaking or pounding the cannabis so the resin glands fall through the pores.

Most Moroccan cannabis plants have a low percentage of resin glands. My advisors told me it took an average of 290 pounds of raw cannabis to make one pound of resin powder. The powder is processed into hashish that sells for less than one dollar per gram in the Rif; in Holland, the best Moroccan hash is nowhere near as strong or as pure as new-method products such as water hash.

Most Moroccan growers have no idea what genetics they plant year after year, nor do they attempt to import seeds from elsewhere that might be better suited to their climate. They're unfamiliar with separating male and female plants to produce sinsemilla. Their processing methods introduce cannabinoid degradation, dust, dirt and hair into the end product.

Further, they were uninterested in new techniques that would improve the quality, quantity, potency, profitability or longevity of their cannabis products.

"We don't need to try new things," one farmer told me. "there are already many tons surplus of hashish and powder stored in Morocco and across the water. We don't mind to grow plants that are not covered with the drug; there is already too much hashish. If we grow bigger and better, we overproduce and the price comes down because of oversupply."

as a dilutant to blend with higher-grade powders. Sometimes inferior powders are mixed with henna, herbs, oils, turpentine and other materials. In England, low-quality adulterated Moroccan and Indian hash powder is sold in toxic hardened bricks called "soapbar."

The best traditional hashish from Morocco, Lebanon, Afghanistan and India smells spicy or piney. The effects of this "double zero" hash

> The best traditional hashish from Morocco smells spicy or piney and the high is dreamy and detached with a mild hallucinogenic overlay.

When processors want more product, they simply beat the cannabis harder and use larger pore sizes, so crushed leaves and other plant material falls through and adds extra volume. This product is darker in color and is inferior the top ranked, pale gold-colored traditional hashish that is often called "double zero," "primero" or "supreme." This secondary hash is often sold in Dutch coffeeshops as "Honey Maroc," or "Golden Maroc."

Some hashish manufacturers literally beat the crap out of their plants so that mounds of pulverized material can be gathered. Such manufacturers often use plants harvested too early, which means the plants have a low mature resin gland percentage. Prematurely harvested plants have not had time to produce a full range of complex cannabinoids.

Material gathered from young or low-grade plants is often used

come on slowly, and the high is dreamy and detached, like that of opium or a sedative-hypnotic prescription drug combined with a mildly hallucinogenic overlay.

## Politics

The government has often tried to stop Morocco's cannabis trade. King Mohammed V, who reigned from 1927 until 1961, sent anti-cannabis military forces into the Rif during the late 1950s. His son, King Hassan II, tried to destroy a large portion of the region's "kif crop" during the early 1960s. There has been an attempt in the government since that time to at least keep the Riffies from accumulating too much wealth or power.

Morocco's kings have allegedly issued royal edicts granting Rif families the right to grow cannabis forever as long as the crop is only used for production

of domestic kif. But the Riffies have been especially paranoid since the Moroccan government assured the United Nations in 2001 that the country would eliminate all internal hashish production by 2008.

Every few years, the Moroccan government makes half-hearted attempts to mess with the Rif, but these efforts only result in cultivators moving their fields up and down the Rif and to the Atlas Mountains south of the Rif, in a game of cat and mouse that government forces never win.

Marijuana cultivation is tolerated because many areas of the Rif are unable to support other industries that would generate the $3 billion per year in foreign exchange and 25,000 jobs that the Rif's cannabis industry creates.

Indeed hashish production is one of Morocco's most successful industries. It is usually ranked above orange groves, date palms, tourism, fisheries and phosphate mining as one of the few capitalist success stories in what is otherwise an impoverished Third World country.

However, Maroc hashish and tourism do not mix together well at this point in time. Morocco is a Muslim kingdom with a harsh government currently infiltrated by U.S. agents. The safest way to experience quality Moroccan hashish is to pay a visit to Spain, or the coffeeshops of the Netherlands.

> Every few years, the Moroccan government makes half-hearted attempts to mess with the Rif, but these efforts only result in cultivators moving their fields . . . in a game of cat and mouse that government forces never win.

# Timing is EVERY

When I first started growing pot in the early 1970's, the relationship between light timing and flowering was virtually unknown by the apprentice grower. The concept of 'bud cycle' was not apparent until Ed Rosenthal and Mel Frank published their first works in 1976. Prior to this, many of us simply grew big plants, either outdoors or under some form of artificial light, and just consumed whatever presented itself.

Young cannabis will flourish practically anywhere, under almost any conditions—but it takes a special environment, and a specific set of circumstances, to properly mature. The key element is the timing of the light cycle. Like humans, plants have two worlds in which they exist: night and day. Put simply, day is when it is light and night is when it is dark.

## Vegetative stage

Sprouts, fresh clones and young plants live in what we refer to as the vegetative stage. This period has a long daytime and a short nighttime, like summer. It is during this vegetative stage that the plants send out much new growth. Large shade leaves form and act as sugar factories for the plant, turning sunlight into fiber for new growth. The plant needs to use as much food from the available light as it can, while it can.

This is why high-nitrogen fertilizers are so beneficial during this period. The nitrogen, coupled with the extra light, acts as building blocks to the overall structure of the plant. As a general rule of thumb, in the indoor garden the average light cycle for the vegetative stage is 18 hours on and 6 hours off.

## Bud cycle shift

At a point in the young plant's development, it becomes time to shift to the flowering cycle. During this period, the plants declare their sex and produce large floral clusters that become the buds. This stage has a shorter daytime and an increased nighttime, such as late summer and fall.

The large shade leaves begin to die and fall off as the plant shifts its energy from producing leaves and stems to producing floral clusters. Food from light and nitrogen decreases, and the demand for phosphorus and potassium increases to fuel the process. During the early flower stage the plant will go through what appears to be a growth spurt as the stems stretch to catch the light that, if outdoors, would be coming at a lower angle as the season progressed.

The floral clusters sprout from the areas where the leaves attach to the stems, called "nodes." The buds fill in the nodes and progress out. And, as we all know, it is strictly the female plants that develop into our high quality and most desired sinsemilla. Indoors, the typical light cycle used in the bud stage is 12 hours on and 12 hours off.

Outdoors, the change in light cycle timing is gradual and slow, a little bit more every day. The transition between the plants' stages is therefore more drawn out and gradual. Indoors, the change in light cycle is usually instant: one day is 18 hours long and the next (and all those to follow) is 12 hours long. Under these indoor conditions the plant is forced to make the shift quickly, which is

# THING

why the average length of the indoor flowering cycle is eight to nine weeks. This forced flowering has its advantages, as the plants are made to finish up quickly, thus aiding production.

Oddly enough, plants do most of their fiber production at night, which may help to explain why such small vegetative plants are capable of producing so much bud in such a relatively short period of time. A well-formed, six-inch-tall veggie plant (a plant in the vegetative stage) placed immediately into the bud cycle, is capable of producing an ounce or two of finished product in two short months, given adequate light, food and root space.

## Sativa variations

Another aspect to consider is that Indica and Sativa varieties differ in their photoperiod expression, or photoreactive rate. The typical 18/6 and 12/12 light cycles are primarily beneficial to Indica varieties. Indica became the herb of choice early on in the industry due to its fast maturation and large production abilities under the HID lights. Indica is a variety from the 30th

parallel and above, and this timing cycle is more akin to locations north or south of the 30th latitude.

Sativa originates from equatorial regions, between 30 degrees north and 30 degrees south. Around the equator there is a much smaller difference between seasonal day lengths. The vegetative stage may be 13 hours of day and 11 hours of night, whereas the flowering cycle may be the opposite, 11 hours of day and 13 hours of night. There are pure Sativa strains that require three to four months to mature in the flowering cycle indoors. And although outdoor equatorial crops take such a long time to mature, it is often possible in the right areas to produce two to four crops per year, thanks to the tropical environments.

Light fortified greenhouses are capable of producing high quality herb just about anywhere on the planet. Once implemented, the global environment will surely reveal interesting and desirable variations, via careful selective breeding. In the meantime, further experimentation and research using different indoor light timing cycles would be very worthwhile. 🍁

## NIGHTS OF TOTAL DARKNESS

An important thing to remember about the indoor bud cycle is that the dark period must be absolute and uninterrupted. The room must be thoroughly sealed to be completely dark when the lights are off. The only way to test this is to sit in the room in the dark, either during the day or with any lights outside the room on, to check for light leaks. It is advisable to allow your eyes to adjust to the darkness before declaring the room adequately sealed.

Once the bud cycle has begun it is important to never interrupt the dark period with any light, even for a short period of time. Doing so may interrupt the long, slow process of change that the plant had been working on up to that point. The plant may react by having to restart the process and seriously delay the scheduled maturation time.

I don't understand why it is that outdoor plants are not as sensitive to these nighttime interruptions. Perhaps it has to do with the unmatchable light intensity of the Sun. Stars, the moon and streetlights glowing through the low clouds over an urban area don't seem to hinder the outdoor plant all that much. For whatever reason, indoor plants tend to be ultra-sensitive to nighttime interruptions of light. So remember to make it dark and keep it dark.

4:19

# by DJ SHORT

# Control your cannabis

**Story by DMT**     **Photos by Barge**

Why waste light and electricity growing stem? There are several ways to reduce internodal length and thus grow denser, more efficient buds.

tight internodes

long internodes

## Temperature control

The easiest and most under-used way to control internodal stretch is temperature control. Plant internodal length is directly related to the difference between day and night temperatures – the warmer your day cycle is as compared to your night cycle, the greater your internode length will be. The opposite also holds true; the closer your day and night temperatures, the shorter your internodes will be. Ever notice how as the warmer summer months approach, your plants begin to stretch? A large factor is the difference between day and night temperatures.

In the grow room, maximum temperatures should ideally never rise above 26°C, so you must do everything you can to prevent your room getting too hot (run lights at night, use exhaust fans, air conditioners, etc). An ideal temperature range is 24-25°C when the lights are on, and 22°C when the lights are off.

The temperature technique is most effective under a 12/12 light regime, which is ideal as this is when cannabis stretches the most. When the light cycle is brought to 12/12 we will raise the night temperature to the daytime level of 24-25°C. Space heaters on timers work well for this, and max/min type thermometers are ideal for tracking temperatures.

It is during the first 2-3 weeks of the flower cycle that most strains begin to lengthen internodes, making it a very important time to control temperature, as this is when the framework for future colas is built. After this 2-3 week window we need to drop the night temperature back down to 22°C, as this is where the plant is happiest.

As floral development begins, the total size of your buds is largely determined by average daily temperature, provided it does not exceed optimal. So if you are letting your day temperatures drop below 24°C or your night drop below 22°C, you are costing yourself in overall weight and harvest.

Once your buds have reached optimal size and and you have begun the flushing period, you may consider dropping temperature down to 17-19°C for the final week or two. This drop in temperature triggers anthocyanin production, which intensifies the color of the floral clusters and makes for a showier bud, especially with "purple" varieties. This final temperature change is not always feasible and can be omitted.

For extreme height control you may even use warmer night temperature than day, but be very careful when running settings like this, as even a zero difference between night and day temperatures will lead to leaf chlorosis (yellowing) after 2-3 weeks.

Some things you will notice while using this technique are a change in the leaf angle, upwards during warm days and downwards during warm nights. There is also chlorosis if this is done for too long. Neither symptom is nutrient related. The plants will return to normal when the temperature is changed back.

## Moisture and conductivity

*leaf chlorosis (yellowing)*

Whether you're growing hydro or in soil, the electrical conductivity (EC) and moisture of your medium are two key elements that should be manipulated to meet your needs. Both of these factors control the same thing; the ability of a plant to uptake water and nutrients from the growth medium. (EC measures the level of fertilizer salts in the water.)

A plant grows by first dividing cells then expanding them, and in order to do this it requires water. By limiting the amount of water available to a plant you limit the expansion of cells. This can work for you by keeping your internodes close together, or against you by limiting bud growth. Both the amount of water you give your plants and the

EC at which you grow them control the uptake of water.

A plant's roots act much like a pump, using osmotic pressure to move water into the plant. In order for this to work there must be a larger concentration of fertilizer salts in the plant's roots than in the soil or hydroponic solution, so when the medium's level of salt rises above the roots', the plant will wilt. Raising the salt level in the medium closer to that which is in the roots limits the water availability just the same as if we had provided less water.

During the vegetative stage we want our plants to form very tight internodes, especially under artificial lighting. By allowing the EC to drop below ideal during this stage we are wasting valuable space growing stem instead of bud. Most marijuana strains are happiest when grown at an EC between 1.5 and 1.8, but different strains have different preferences. Try growing one of your plants using straight water for a week or so, you will see the internode length stretch dramatically compared to the ones on a regular fertilizer regime.

Hydroponic tomato growers sometimes grow transplants at extremely high EC's (up to 6 EC!) in order to get stocky production plants. Please note that they use special nutrient formulas designed for this purpose, most of which have potassium to nitrogen ratios of 4:1, much higher than normal, as too much nitrate at this high an EC will easily damage a plant.

With cannabis, I would not recommend going above 3 or 4 EC. Do not try on all of your plants at once until you are sure your strain can handle it. Be sure to bring your EC down once you enter floral stage. By the time tufts of pistils are visible you want to be at your ideal EC of 1.5-1.8.

Try not to change the EC too quickly as a sharp drop can cause root damage. This also goes during your final flushing period when you want to eliminate all fertilizer from the medium — lower the EC over a couple of days, as the sudden change in salt level will harm the roots.

When growing hydroponically, the only way of manipulating water availability is with the EC, while in soil we may also use the moisture level of the medium to the same ends. Many growers are under the mistaken impression that the EC and pH of their nutrient solution remains the same when applied to the soil. This in not the case, and you must test the soil in order to have a true picture.

To test your soil, take a sample from the center of the root zone at the side of the pot. Mix the soil with 2 equal parts distilled water and let sit for 20 minutes. Then take an EC reading and multiply this number by 2.4 (this takes into account the dilution and the pore space factor) this will give you an accurate picture of the EC the roots are actually

being exposed to. The pH should also be checked at this time. It is not feasible in soil to maintain an exact EC at all times, what we need to try and avoid is EC's climbing much above what we want and plants going for long periods with very low EC's.

A frequent mistake marijuana growers make is over-emphasizing the need for a plant grown in soil to dry out completely between waterings. Cannabis does like dry feet but this simply means that the root zone must not be kept extremely wet at all times. Keep in mind that if the soil has an EC of 1.8 and then dries out completely the amount of salt remains the same, causing the EC to double or more.

As a general rule, during the vegetative stage you should keep your plants a little on the drier side as this will restrict cell elongation, creating a shorter noded plant structure capable of creating a dense bud cluster in the floral stage. (Unless of course you are using the high EC method described above, in this case you must not let your soil get too dry because of the increased fertilizer level you will create.) Maintain this level of moisture into the first 14 to 20 days of 12/12 to minimize internode stretch.

As soon as early flowering begins you need to increase soil moisture to a nice evenly moist (not soaked) level to maximize bud expansion. Growing marijuana too dry during this stage will adversely affect your overall yield, as will having too high an EC in the medium.

In these times of government oppression we must make the most efficient use of available growing space in order to produce the copious amounts of cannabis necessary to overflow the boundaries placed upon us. Control your cannabis!

# CANNIBIS CULTURE: RUSSIA

## FROM SMALL BEGINNINGS, SOVIET TOKERS ARE GROWING INTO A POTENT CULTURAL FORCE

### BY ANASTASIA V

Until the 1970s, marijuana was not used much in Russia. Alcohol was Russia's national drug, and ganja was only used in the "wild" Asian republics. Average Russians considered ganja smoking as something alien and strange.

The situation changed during the 1970s, as the echo of the hippie movement came to the USSR, and the immigration of Asians into Russia increased. Pot became popular among Russian "bohemia," providing the chance to stand out against a gray background of gloomy compatriots.

Marijuana remained a bohemian drug for a long time, but in the last decade, it has seen skyrocketing popularity. In 1992 most young people knew what "grass" was, and everybody was acquainted with a couple of tokers. By 1995 almost all of the participants of my extempore polls had inhaled marijuana at least once. Today it's very difficult to find a young Russian who has never smoked ganja.

### POT PEOPLE

Nowadays alcohol is still the most popular drug among Russians, but more young people are discarding alcohol in favor of marijuana.

Although pot smokers all over the world have many of the same smoking rituals, there are some specific features of pot smoking in Russia worthy of mention.

To begin with, Russians very rarely roll joints. They have excellent *papiroses*—short thick (tobacco) cigarettes without filters. There are many different brands of *papiroses* in Russia, but Belomorcanal or simply Belomor is the most popular. However it's usually preferred by low-paid industrial workers, not well-dressed young people.

*Papiroses* are either filled with pure pot or mixed with tobacco. The *papirose* filled entirely with weed is usually called *kosyak,* a word that means "jamb" or "shoal."

Friends usually smoke a *kosyak* using a technique called *paravoz,* which means "steam engine." *Paravoz* is quite a ritualized method of pot smoking. There are two participants. One takes the burning end of the *kosyak* inside his mouth and breathes out. He is said to be "giving out *paravoz.*" The other draws a deep breath from the opposite side of the *kosyak,* and is "receiving *paravoz.*"

*Paravoz* is a collective ritual that is popular in big gatherings. In spite of its outward eroticism, *paravoz* is used irrespective of sexual orientation.

Sometimes marijuana is smoked through a *burbulyator* (an untranslatable word imitating the gurgle of boiling water). A *burbulyator* is a bong-like device made from a plastic bottle. The bottleneck is cut off and put into the bottle. This part can now move up and down. The bottle is filled with water, and the upper part is moved down. The bottle's neck is covered by aluminum foil, pierced and filled by the grass. A smoker lights up the grass and slowly raises the upper part of the bottle. Smoke appears in the bottle. Then the toker takes off the cap and moves down the upper part of the bottle, inhaling the chilled smoke going out of the bottle through the neck.

Particularly in the southern regions of Russia, there is a very popular pot dish known as "milk." It is made of grass boiled with milk fat and butter. The milk is a greenish drink with an astringent taste. If made properly, it has strong psychoactive effects with intense visuals. However, it requires a quantity of grass to prepare, so good pot is usually smoked. A proverb says that weak grass "can be used only for milk."

### RUSSIAN LAWS

In spite of stories about their brutality, Russian cops are not so dangerous as their colleagues in parts of the U.S. Yes, they are rude and used to solving all problems by brute force. But they have scanty earnings and most of them would prefer to take a bribe and let a toker go (even with his pot!) rather than bust him without any profit.

How much do you have to give to a Russian cop? It depends on the contents of your wallet. The process of deliverance from a cop is called *otmazka*. It's always

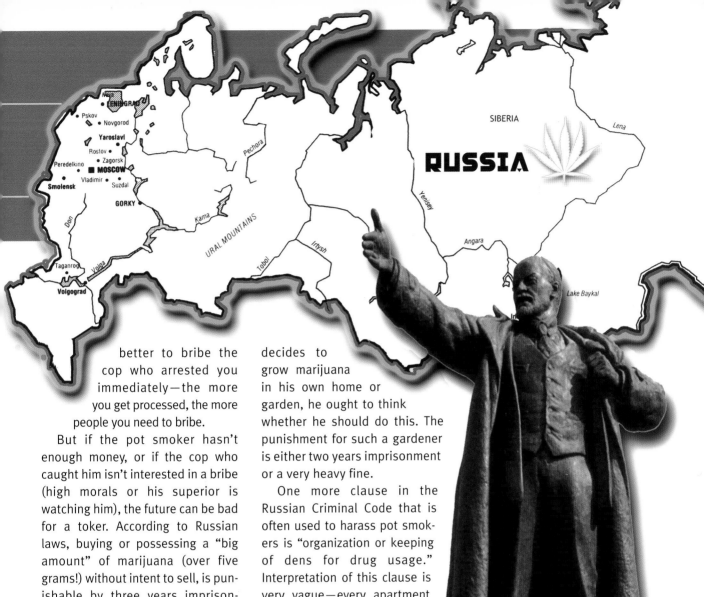

better to bribe the cop who arrested you immediately—the more you get processed, the more people you need to bribe.

But if the pot smoker hasn't enough money, or if the cop who caught him isn't interested in a bribe (high morals or his superior is watching him), the future can be bad for a toker. According to Russian laws, buying or possessing a "big amount" of marijuana (over five grams!) without intent to sell, is punishable by three years imprisonment. If you are convicted of buying or possessing with intent to sell, you can be imprisoned for between three and seven years. Russian law also encourages snitching, which can lighten a sentence.

In addition, there is a clause in the Russian Criminal Code that can be applied to many situations. It is an "inclination to usage of drugs." If you persuade somebody to smoke a bit of ganja, you become a criminal. You can be imprisoned for between two and five years. If several people commit this "crime," they can be imprisoned for a term of three to eight years. The same punishment awaits you if your "victim" wasn't of legal age.

Before the Russian ganja man decides to grow marijuana in his own home or garden, he ought to think whether he should do this. The punishment for such a gardener is either two years imprisonment or a very heavy fine.

One more clause in the Russian Criminal Code that is often used to harass pot smokers is "organization or keeping of dens for drug usage." Interpretation of this clause is very vague—every apartment where people smoke pot from time to time can be called a "den for drug usage." For this "crime," the apartment owner can be imprisoned for a term of three to seven years.

It might seem that these periods of imprisonment aren't very long compared to the U.S., but unfortunately, conditions in Russian jails are quite terrible. Even a short period spent there can be deeply shocking.

## LEGALIZATION

The most widespread thesis of the prohibitionists is that marijuana is a springboard to

*Left to Right:*
- *Belomor* papirose *becoming* kosyaks.
- *Giving and receiving* paravoz.
- *Smoking through a* burbulyator.

## POT WORDS

Russian pot smokers have a voluminous glossary of pot words and expressions. Here are some of the most widespread terms used:

### WORDS FOR CANNABIS

*anasha, plan, konoplya* (hemp), *shala, ganjubas, cannabis, hash, soloma* (straw), *dudki* (pipes), *drap, hashish, plastilin* (plasticine), *matsanka, pyl'* (dust), *shishki* (cones), *boshki* (heads) and *dur'* (folly).

### AMOUNTS

*korabl:* this is a matchbox amount of grass, which is a common way to measure. The word literally means "ship." Ten ships make one "glass."

*pakavan:* a small paper package that people often carry pot in.

### SMOKING

*papirose:* short, thick filterless tobacco cigarette.

*kosyak:* a papirose filled with grass.

*pyatka:* literally the heel; it is one-third to one-quarter of a whole kosyak; an amount of the kosyak that might be smoked if the pot is strong.

*shtaket:* an empty papirose.

*zabit:* to stuff a papirose. It literally means "to choke up."

*prikolotit:* also to stuff a papirose. It literally means "to nail."

### EFFECTS

There are many words for the effects of marijuana, too. Its primary effects are designated by the words *vstavlyaet* (puts in) or *vpiraet*. When the grass is weak *(bespontovaya)* Russians say that "it doesn't put in," and you have to "catch up yourself" *(dogonyatsa)*. If the grass is strong, it "puts in" after a couple of draws.

*"Mne vperlo," " nme vstavilo,"* or *"menya pryot"* is equivalent to "I'm high" in English. For the intensification of emotional effect of the words, you would add some epithet, such as *"ne po-detski"* ("not childishly,") meaning very strong, and say : *"menya pryot ne po-detski."*

"Pot paranoia," is called *izmena* (treason) in Russian. To "set down on a treason" (*sest' na izmenu*) means to be deeply frightened of something or somebody, both apropos of something and apropos of nothing. Russian pot smokers often "set down on a treason" at the sigh of *ment* (a cop) because many Russian ganja people smoke outdoors and have some grass in their pockets as well.

stronger drugs. "All junkies smoked marijuana before they became heroin addicts. Hence all marijuana smokers will go to heroin hell"— declare prohibitionists without any doubts in their logic.

Anti-drug hysteria reigns in Russia. Newspapers write colorfully about instant addiction, blurring the lines between marijuana and injectable drugs. Intimidated parents are panicked. Teenagers, having tried ganja, learn that the media claims are exaggerated. "If marijuana is 'drugs,' then drugs are interesting and safe," thinks the teenager brought up by intimidated parents and ignorant journalists. By mentioning marijuana in the same context as heroin, the state itself pushed youth to use other stronger, more dangerous substances.

Legalization, sadly, is not in Russia's immediate future. Before that can happen, it is necessary to educate the Russian people and increase awareness of the cannabis culture in the media, persistently and patiently. 🌿

# LET THEM BREATHE!

BY DJ SHORT

All Photos: Ed Rosenthal

## PROPER VENTILATION AND AIR CIRCULATION ARE ESSENTIAL TO GROWING HEALTHY, HAPPY PLANTS.

*An important aspect to consider when growing plants indoors is the air. Proper ventilation, air circulation and temperature control become especially important when working with lights over 400 watts, in very small spaces, any time the temperature exceeds 32°C (90°F), or if the humidity gets too high.*

### THE PERFECT TEMPERATURE

Although the "feel" is adequate to gauge the "perfect climate" for a given plant, there is no real substitute for a thermometer and humidity gauge.

Thermometers are cheap and accurate enough for our purposes. I usually employ several thermometers in different areas in and around the grow room. Somewhere between 32-35°C (90-95°F) is the absolute highest room temperature your plants would care to tolerate. The perfect temperature would be somewhere between 24-29°C (75-85°F). Peaks of 38°C (100°F) are allowable for most strains, but not for any longer than a half hour or so. And only above the root level.

### ROOTS AND AERATION

The main area of concern involving temperature are the plant roots. Ideally, the roots should be kept at as constant a temperature as possible: below 21°C and above 10°C (70-50°F). The fact that warm air rises and cool air sinks works to our advantage in this case. Also, the plants end up under the larger lights by the flowering cycle, and so they're usually large enough to help shade and cool their root areas.

Still, some rooms build up sufficient heat to require a separate circulating fan, or fans, focused specifically on the root systems. A soil thermometer may be a wise investment.

Proper aeration of organic based soils is crucial in high temp/humidity areas. Perlite and vermiculite are the tips here – add more to the soil if need be. In hydroponic systems, make sure that the nutrient water temp is below 21°C (70°F). If necessary, store the reservoir outside of or below the grow room.

### SQUIRREL CAGE AND HOUSE FANS

There are many different types of fans and air movers available on the market. Most fans can be purchased at the average home improvement store. Proper research and smart shopping will net the best purchases. Careful planning will help avoid costly mistakes. Using the "hot air rises, cool air drops" rule, one can figure out the right solution.

The two most common types of fan are the squirrel-cage and what I call the "common house fan" (box or oscillating fans). Both come in a seemingly endless variety of shapes and sizes.

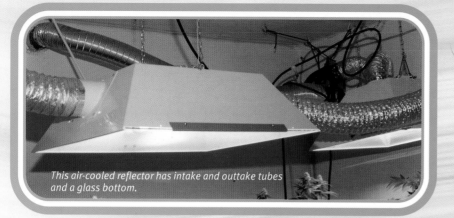
*This air-cooled reflector has intake and outtake tubes and a glass bottom.*

Generally speaking, squirrel-cage fans move air either in and/or out of the room, while common house fans move the air up, down and around the room. There are also neat little "muffin" fans that can be used for many things such as light-hood venting and passive ventilation systems.

A passive ventilation system is one that moves air either in or out of the room (not both). The room is not sealed, so air exchange is allowed free movement from inside and outside the room.

## SQUIRREL CAGES

The squirrel-cage fan is the most popular fan for moving large volumes of air into or out of a room or rooms. A common placement for this fan is inside of the room, up high, blowing out. This will help move the hot air out. This method is what is used to stimulate the passive intake of cool air with vent holes cut in the floor or lower walls to access the cooler areas outside of the room.

Other hardware such as dryer-vent tubing or muffin fans may be used to best access the cool, dry air outside of the grow room. It is a simple step further in this type of system to add an oscillating fan or two on the floor, pointing at any angle up, to help circulate the cooler air up and around the plants. This is the simplest of vent systems and works quite well.

Choosing the correct squirrel-cage fan is part of the trick to success.

## MEASURING AIR MOVEMENT

Squirrel cage fans are rated by their volume of air movement in cubic feet per minute or CFM. A fan with a rating of 100 CFM is able to move 100 cubic feet of air per minute. A room that is eight by ten feet and eight feet tall holds 8 x 10 x 8, or 640 cubic feet of air. Therefore, it would take an optimally running 100 CFM fan 6.4 minutes to fully circulate the air in that room.

Generally speaking, most fans move a little less than their rated CFM due to intake resistance or a dirty fan cage. Bigger fans usually will work more efficiently. Potentiometers, or a "volume control," could be installed in the power line of the larger fans to adjust the fan speed. This would

give further aid in the specific control of air volume and ventilation.

## AUTOMATION

The ideal ventilation system utilizes automtion in the form of thermostats and regulators. A thermostat, as with the common household thermostat, would cause the fans to turn on at a certain temp, and turn off at another. That is, a sensor would turn on the fans on at around 30ºC (86ºF), and turn them off if the temperature dropped below 21ºC (70ºF). A well-stocked, high-tech grow shop will have several types of thermostats available in a variety of systems.

## BOX AND OSCILLATING

Common summer house fans also come in a wide array of types and sizes. The most common being the box and the oscillating. Box fans are self-explanatory. They can be used in a variety of ways, depending on the innovation and imagination of the user. Experimentation will yield the most efficient uses for these devices.

Oscillating fans are perhaps the most efficient devices for circulating air in a room. The gentle back and forth sway of the fan is

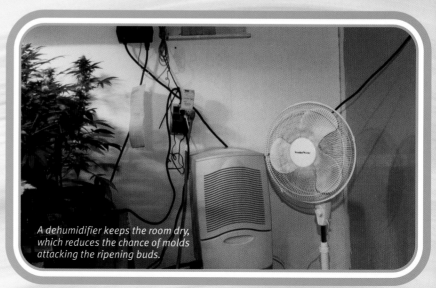
*A dehumidifier keeps the room dry, which reduces the chance of molds attacking the ripening buds.*

*Oscillating fans keep the air circulating in the garden. Mylar curtains (in the background) keep the light from wandering out of the garden.*

very beneficial for the developing plants. These fans tend to keep anaerobic molds down by constantly freshening any potentially stagnant air. There are wall-mounted styles available as well. Home improvement centers carry a large array of various types and styles of air-moving fans, some relatively inexpensive.

A warning needs to be expressed concerning the cheaper, discount-store, oscillating fans (or any cheap fan for that matter) that have a tendency to burn out after a period of time. Some of these products are potentially dangerous if left plugged in and turned on after they burn out. Therefore, it is a wise idea to check one's fans (and all electric devices and equipment for that matter) on a regular basis.

## NOISE CONCERNS

Another fan consideration is noise. Some fans, especially the squirrel cage, may be a bit too noisy for a given situation. There are higher quality fans available that do run more quietly – expect to pay more, of course. It also

helps to mount the fan directly to a main stud or support, by at least two of its support holes, and preferably more.

Rubber dampers and gaskets can be easily made and used on the support holes or around the overall mounting surface. Keep the fan's bearings sufficiently lubricated as well.

## HIGH HUMIDITY

Humidity is another factor that influences the overall quality and quantity of a crop. Generally speaking, high humidity (over 80 or 90%) is bad. It inhibits plant transpiration and ultimately stunts growth. Mold and fungus love high humidity as well. Note that warmer air holds more moisture than cooler air.

There are a few simple practices to help reduce humidity. First and foremost, keep the room as dry as possible. When watering, use just as much as the plants need. Pump, siphon or mop up any remaining water and remove it from the room.

Keeping the room clean also helps. Moisture likes to hide and store itself in material such as dead leaves, spilled dirt or any garbage. Therefore, keeping the

room clean and free of debris will help keep moisture and organisms such as mold, fungus and bacteria down.

Temperature and moisture levels directly affect the plant's ability to metabolize nutrients and supplements such as fertilizer and carbon dioxide.

If these practices fail to lower humidity enough, the only solution may be a dehumidifier. However, dehumidifiers are expensive, consume a large amount of electricity and produce heat. These factors will need to be considered in choosing whether or not to employ one.

## VENTILATION AND CIRCULATION ARE ESSENTIAL

Proper ventilation and air circulation are essential to maintaining a healthy indoor grow environment. The basic rule of thumb is to move the warm, moist air out and to move the cool, dry air in and around the plants and their roots. Many various types of fans and devices are available to achieve this goal.

Careful planning, basic research and smart shopping will acquire all that one needs to keep it cool and dry, and experimentation will fine tune the system to provide the most perfect indoor environment possible. 🌰

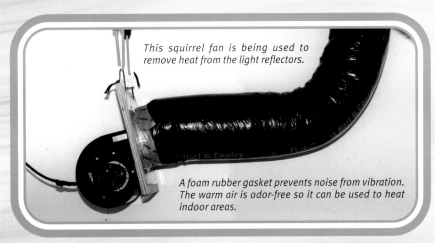

*This squirrel fan is being used to remove heat from the light reflectors.*

*A foam rubber gasket prevents noise from vibration. The warm air is odor-free so it can be used to heat indoor areas.*

photos by Ed Rosenthal

# POT POTENCY

Marijuana is perhaps the only plant in history grown primarily for its trichomes. Whether used for hashish or marijuana the objective is the same: production and harvest of THC-rich trichomes (crystals). Yet often the cultivator loses sight of the real goal of growing this plant. Bud size, density, coloration, flavor and odor are all of little value if they do not produce the euphoric sensations we seek.

## FORGET BIG COLAS AND STINKY BUDS: IT'S ALL ABOUT THE RESIN.

### TRICHOMES AND RESINS

Trichomes come in many shapes and sizes and are used by plants for many different purposes. Cannabis uses its trichomes for a variety of purposes, some of which require THC and other cannabinoids to be effective, and others that do not.

The primary goal of any plant is to create and nurture seeds to the point where they will be viable for future growth. Trichomes help prevent seed damage from desiccation, insects, animals, light degradation and fungal disease. Perhaps the most successful function of trichomes in the proliferation of cannabis is their attractiveness to humans. What better a creature to protect and spread a genus than the most advanced organism on the planet?

photo by Tom Flowers

STORY BY DMT

An important thing to remember is that heavy trichome production is not necessarily an indication of a potent plant. Some hemp strains

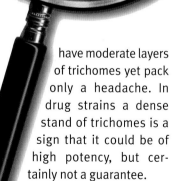

*Photo: Barge*

have moderate layers of trichomes yet pack only a headache. In drug strains a dense stand of trichomes is a sign that it could be of high potency, but certainly not a guarantee.

This is because the resins that flow within may or may not hold the THC and other cannabinoids that we are looking for.

Indica varieties often look more heavily crystallized than Sativas, yet typically don't have the same mind-warp capabilities.

Even with a known high-THC clone, THC level and cannabinoid ratios may change depending on environmental conditions.

What defines drug strain cannabis is the plant's ability to convert cannabidiol (CBD) or possibly cannabichromene (CBC) into THC.[1] If we as growers do not provide the plant with reason to make this conversion, it likely will devote its energy elsewhere to aid in its survival.

## ENVIRONMENTAL INFLUENCE

It takes high quality genetics to produce high quality marijuana, but genetics is only half of the equation. The genetic structure (genotype) only plays 50% of the role in determining the appearance and quality (phenotype) of a given plant. The other half is determined by environmental conditions such as light, temperature, humidity and soil nutrition. All these factors play a role in both the physical and chemical nature of marijuana's trichomes.

The best way to take a look at how environment affects THC production is to look where on the planet cannabis has naturally adopted a high THC profile. As cannabis has spread around the world, it has taken on many different traits to help in its adaptation to varied areas. The best drug varieties have always been found at equatorial or high altitude locations. The one thing which both of these variables have in common is high light intensity and a large amount of ultraviolet (UV) light in the spectrum.

Recent Swiss trials in outdoor plots of clones grown at different altitudes have shown that there is correlation between higher altitude and increased potency (although there seems to be a trade off in yield). This likely means that THC-rich resins act to protect the plant and its seed from both higher light intensities and ultraviolet presence. It's no surprise that cannabis has developed a chemical to protect itself against the sun's damaging UV rays, as they can be injurious to all forms of life.

In a plant's search for survival, energy put towards unneeded processes is wasted energy. Therefore a high-THC plant grown in a low-THC environment will likely produce a medium THC result.

Humidity also plays a role in plant resin production. Although some potent equatorial strains do seem to occur in high humidity areas, most high-test land races have evolved in drier areas, like Afghanistan. The aridity of the areas of Afghanistan where Indica strains have evolved is quite apparent by the trait of large dense flower clusters. This would only be an advantage in an area of low humidity, as flowers will mold in anything more.[2]

There are many examples of non-cannabis plants producing resins in order to protect themselves from drying out. The waxy coating on cacti and other succulent plants is a prime example.[3] Marijuana that is flowered in humid conditions often has a longer stalk on the glandular trichome than the same strain grown in drier conditions. While this may give the appearance of being very crystallized, it probably contain less THC than the same plant grown in a drier environment. Another problem with longer trichome stalks is that the gland heads are more likely to break off during handling.

Shiva Skunk: Your magnifying glass is your friend.

### FLUSHING: PROS AND CONS

Much time and thought has been put into the feeding needs of each part of marijuana's life cycle, yet for some reason the final stages of resin development always seem to be ignored. But the vegetative period of plant growth is only setting the platform for us to produce the trichomes that we are after.

Flushing in particular seems to be something that is over-emphasized by many of today's growers. Many growers "flush" their plants with straight water or clearing agents during the final weeks before harvest in an effort to improve taste and smokeability. The theory is that this forces the plant to use up stored nutrients that may affect these qualities. Although this is certainly true to some extent, what many are forgetting is that not all nutrients can be moved within the plant.

Nitrogen, which is the main factor in poor-tasting bud, can be moved within the plant. If not present in the root zone a plant will take it from the older leaves to support newer growth. Calcium, however, is a nutrient that cannot be moved within the plant, if it is not present in the root zone it is not available for growth. Little research has been done on nutritional requirements of cannabis during the final stages of flowering, but it seems calcium could be vital, as it is crucial in cell division. A calcium deficiency at later stages could therefore adversely affect trichome production.

This is not as serious of a concern for soil-based growers, as lime or other calcium sources which are mixed into the soil will provide sufficient nutrition even while flushing with pure water. But hydroponic growers using very pure water sources with little naturally occurring calcium could have problems. Flushing is certainly a valid technique, but is easily

Photo: Tom Flowers

overdone and is not a quick fix for overfeeding earlier in the flower stage.

Some studies have shown that high potassium levels have a negative influence on THC production,[4] which would correlate to the general belief that while hemp crops uptake more potassium than phosphorous, the reverse seems to be true for drug and seed cannabis crops.[2] A study on how to minimize THC levels in hemp crops showed that THC levels in newer leaf growth decreased as nitrogen levels were increased.[5] As no THC measurement was taken from floral clusters, we can only speculate that the same would hold true in buds. This would also explain the good results that most growers have flushing their plants, as nitrogen is the nutrient most easily flushed from the soil.

### COMPANION PLANTING

Much research is still needed on the interrelationships of plants in the garden. Little is known about common vegetable garden plants' effect on each other, let alone how they may react with cannabis.

Growing certain plants in proximity to each other has been documented to cause noticeable effects on growth, both positive and negative. The main companion plant that has attracted interest with underground marijuana researchers is stinging nettle (Urtica dioica) which has been said to increase essential oils in many plants.[6]

### BREEDING FOR POTENCY

Marijuana is unique from an evolutionary standpoint. It is the only plant in history that in some cases has been grown and bred for over two decades under nothing but artificial light. It is very likely that there have already been some genetic changes that have taken place as a result of this. All plants, especially cannabis, will quickly adapt to a new habitat by adding or dropping traits over successive generations. With breeders doing potentially as many as three or four generations per year, over 20 years there is great opportunity for drift from original genotypes.

Some "oldtimers" of the cannabis community have theorized that the use of high pressure sodium (HPS) light as a sole source of lighting has resulted in unconscious selection for lower THC parents during breeding.[7] This theory is based on the assumption that ultraviolet light is a large causal factor in the plants' production of THC. As HPS lights produce little in the way of UV, the lower potency plants could look the most vigorous in early selections (before flowering) as they would have a genetic advantage over high THC plants (less wasted energy).

A common way of conducting a breeding program where space is limited is to start large seed lots and then select the best individuals for flowering. Vic High and others have done some preliminary research into creating high-UV environments by adding tanning or medical UV lights to their regular lighting for early seedling selections.[8] As most Dutch breeding is done behind closed doors, it is unknown whether this is used by any breeders in Holland.

Photo: Barge

Kali Mist: Sure, it looks good, but does it have phat trichomes?

## TRICKS OF THE TRADE

Delving through the history of marijuana cultivation, you will find a myriad of techniques used to supposedly increase THC production. Much of this is little more than hippie folklore, but over the years some techniques have appeared which seem based on some amount of science.

Although some of the younger growers these days may never have used a metal halide light, many of the older set still swear by them as a complement to high-pressure sodiums in the flower room. With the advent several years ago of the Son-Agro HPS bulbs and others like it, which offer a higher amount of blue in the spectrum than standard HPS, many growers have felt that that they can do away with metal halides altogether. Growing strictly under sodiums has its advantages in terms of yield per watt, but is still lacking as far as a balanced spectrum when compared to a mix of HPS and halide.

Anyone that has ever seen a mixed light garden can testify that the healthiest, most crystallized buds occur where the two spectrums overlap. Again this brings us back to the UV factor, as metal halide bulbs emit a fair amount of UV while HPS emit almost none. Most growers employing halides in conjunction with HPS do so at a 2:1 HPS:MH ratio. Many growers, especially those restricted to one light, have been having good success using one of the new enhanced metal halide bulbs such as Sunmaster, which have a more balanced spectrum than either sodium or regular halide alone.

Glass and plastic materials used in greenhouses and air- or water-cooled light reflectors will block most useful wavelengths of UV from reaching plants. Luckily, recent research has shown that allowing UV to enter the greenhouse has many advantages on non-cannabis crops, and so some European greenhouses are beginning to switch to UV transparent glazing materials. Trade names for some of these products are Planilux, Diamant or Optiwhite. Plastic made from polymethyl-methacrylate (PMMA) also transmits UV-B (the type that we are looking for). Traditional greenhouse coverings such as polyvinyl chloride (PVC), fiberglass, polycarbonate or regular glass allow little if any UV-B transmission.[9]

Harvesting in the morning ensures that your plant will be at peak THC content, as cannabis has shown THC fluctuations peaking in morning and dropping during the day. Some growers leave their lights off for several days before harvest to increase potency. This seems to have some scientific validity as light has been shown to degrade THC, hence the morning peaks. As light is the degrading factor and the plant still has the ability to manufacture THC during darkness, leaving the lights off for a day or two before harvest likely utilizes the plants stored potential for THC conversion without any opportunity for it to be degraded into cannabinol (CBN) and other breakdown products.[8]

Traditionally marijuana has been harvested when the pistils die and the calyx starts to swell into a false seed pod. These days the best growers are getting much more detailed in their harvesting criteria. They take a close look at the trichomes themselves to judge peak harvest. Evidence that this is the only real way to tell peak maturity is in Sagarmatha's strain

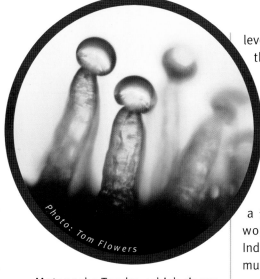

Photo: Tom Flowers

Matanuska Tundra, which ripens resin glands while most pistils are still alive and white. This seems an odd twist of evolution but proves that the pistil color and ripe glands do not necessarily have any correlation.

A small 25x or more pocket microscope, which can be picked up inexpensively at any electronics store, works well for taking a closer look at trichome development. What we are examining are the capitate stalked glandular trichomes, which will be a round gland head supported on a stalk. The coloration of these gland heads can vary between strains and maturity. Most strains start with clear or slightly amber heads which gradually become opaque when THC levels have peaked and are beginning to degrade. Regardless of the initial colour of the resin head, with careful observation you should be able to see a change in coloration as maturity levels off.

Some cultivators wait for about half of the heads to go opaque before harvest to ensure maximum THC levels in the bud. Of course nothing tells the truth more than your own head, so try samples at various stages to see what's right for you. While you may be increasing the total THC level in the bud by allowing half of the glands to go opaque, the bud will also have a larger proportion of THC breakdown products such as CBNs, which is why some people prefer to harvest earlier while most of the heads are still clear.

Indica varieties usually have a 1-2 week harvest window to work with, while Sativas and Indica/ Sativa hybrids may have a much longer period to play with.

## GLANDULAR CONCLUSION

With the growing popularity of personal hashmaking through precision screening, many growers are starting to pay closer attention to the development of glands. The use of different gauge screens to separate glands of different sizes can only broaden our knowledge of the subtle nuances of trichome quality.

Growers using the same clone line over many crops have an excellent opportunity to play with some of these different techniques, as the main variable will be the environment, not the plant. Keep in mind that different strains may react very differently to the same techniques so be careful about drawing general conclusions.

Marijuana growers must look closer at their crop than the average farmer to achieve a premium product. Rows upon rows of beautiful plants are of no use if they do not glisten with the THC-laden trichomes that are the object of our quest.

Nurture your trichomes and feed your head! 🍂

REFERENCES:

1) Starks, Michael. 1977. *Marijuana Chemistry Genetics, Processing and Potency*. Ronin Publishing, Inc., Berkeley, CA pp. 17-86.

2) McParland, Clarke, Watson. *Hemp Diseases and Pests: management and biological control*, CABI Publishing, New York, NY

3) Pate, DW, 1994. *Chemical ecology of Cannabis*. Journal of the International Hemp Association 2: 29, 32-37.

4) Kutscheid, 1973. *Quantitative variation in chemical constituents of marihuana from stands of naturalized Cannabis sativa L. in east central Illinois*. Economic Botany 27: 193-203.

5) Bócsa, Máthé and Hangyel. *Effect of nitrogen on tetrahydrocannabinol (THC) content in hemp leaves at different positions*. 1997. Journal of the International Hemp Association 4(2): 78 -79.

6) Helen Philbrick and Richard B Gregg. *Companion Plants and How to Use Them*. 1996. Devin-Adair Company, Old Greenwich, CT.

7) Oldtimer1, 2001. Personal communication

8) Vic High, 2001. BC Growers Association. Web site and help desk.

9) Hoffman, Dr Silke. 2001. *Ultraviolet radiation in the greenhouse*. Floraculture International, May 2001. Ball Publishing, Batavia, Illinois. pp18-27.

• An excellent general reference is *Marijuana Botany*, by Robert Connell Clarke. Ronin Publishing, Inc. Berkeley, CA

# Love You

**Healthy mother plants are the key to any succesful grow operation. Yet unfortunately they are often being held a distant second to the flowering clones. Just as it is impossible to grow the highest qualit**

*left: this special mother has provided many clones, but has been over-pruned.*

*right: trays of new clones grow under the watchful eye of their mothers.*

## STARTING FRESH

Marijuana is an annual plant; it completes its life cycle in less than a year. Nature's way of keeping diseases and viruses at bay is to start fresh from seed each year. Starting with seed also helps prevent loss of vigor and other pitfalls associated with keeping clone lines around for a long time.

Any serious grower should really try to start their own mothers from seed. When purchasing seed, you get what you pay for. Small independent breeders are great for specialty items, but simply do not have the large grow-out space that gives the consistent results of seed giants like Sensi Seeds. One factor often overlooked in the search for a perfect mother is that even the most stable seedlines show variance, and therefore the larger number of plants you grow from seed to select from, the better your chances of finding that really special mother.

The selection you will get from your pack of 10 seeds cannot compare to a huge greenhouse operation. So always grow out as many plants from seed as possible, ensuring that you are off to the best start.

## VIRAL THREATS

Viruses are a threat that many growers remain unaware of. They are usually spread by insect feeding, and can wreak devastation in a clone garden. Viruses often sit latent until ideal conditions present themselves, at which point they can produce symptoms ranging from slight streaking to induced nutrient deficiency or full plant collapse.

In North America the main vector of viruses are thrips. If your mothers have ever had thrips problems in the past, they will likely be carrying a virus which will be passed down to any clonal generations. Thrips will usually be found hiding in the growing tip of

the plant; they are long and slender and can be seen with the naked eye if you look close enough.

If you have any flowering male plants in the vicinity, this is where you will find the majority of the thrips, as they enjoy feeding on pollen. Thrips produce a sticky honeydew on the leaf surface which often turns black from a mold that grows on it. They also leave feeding marks on the leaves.

## LIGHT AND CUTTINGS

Strong lighting and short plant height are necessary to produce strong cuttings. Cutting maturity and stem diameter play a large role in producing quality, high yielding finished plants. Spindly cuttings taken from the bottom of a plant are a waste of valuable production space.

Mother plants should have the main growing shoot pinched, forcing the plant to branch out and create more cutting sites. Keep the plant trimmed to maintain no more

# Mother

photos by Barge

, grown under insufficient lighting, continually stressed from being stripped of cuttings and generally s from poor genetics, clones taken from an inferior mother will produce poor quality bud.

than a two foot deep canopy, this will allow maximum light pentration, resulting in strong cuttings.

When taking cuttings, only take from strong lead branches, being careful to sterilize the knife between plants.

## PH AND NUTRIENTS

A common problem when growing mother plants in soilless mix is a gradual lowering of the pH in the medium. This is primarily caused by constant use of veg-cycle fertilizers, in which the ammonium nitrate, urea and phosphorus acidify the soil over time. A quick check may be done by mixing some of the soil with a similar amount of distilled water, letting it sit for 20 minutes, then checking with a test strip. The pH should be no lower than than 5.8; if it slips below this point either use pH up, hydrated lime or potassium nitrate at an EC of no more than 1.8.

Additions of magnesium may also be needed periodically in order to keep optimum quantities in the plant. One feeding of magnesium sulphate (Epsom salts) at 1.8 EC every three or so weeks is sufficient.

Non-hydroponic fertilizers such as 20-20-20 are especially problematic because they contain little or no calcium or magnesium, as these react with phosphorus when concentrated. A and B type formulas eliminate this problem.

## A HAPPY MOM

Take good care of your mothers and they will pay you back with resin-laden children of bountiful proportions. The more serious a grower you are, the more important these steps are for you to take. With a happy and healthy mother, your cuttings will thrive and bear bountiful buds! 🐛

## Insect control

Careful monitoring for insects and starting with clean plants cannot be over-emphasized. Your mothers must be completely insect-free to produce successful cuttings. A well orchestrated insect control program must be implemented from many angles.

### SCREENED INTAKE

To prevent insects from entering your grow show, any intake openings should be screened. The maximum hole size to exclude thrips is 192 microns, and they are the smallest pest we need to worry about.

Mesh of this size will seriously restrict airflow. Compensate by creating a larger surface area of screen to pull air through. Properly sized silk or other printing screen works well for this application.

### SAFE SPRAYS

Monitoring should be carried out both by visible inspection and sticky cards placed throughout both the mother and flower rooms. In the veg room no tolerance for insects is acceptable. At first sign, the entire room should be sprayed with the appropriate insecticide.

At normal grow room temperatures, thrips must be sprayed 3 times, once every 5 days. This allows young thrips to develop into adults and be killed before they can reproduce. Spider mite sprays should be spaced a little further apart, approximately 7 days.

*A happy family: mothers and clones growing together.*

Whenevever possible use earth- and people-friendly insecticides such as neem and cinnamon-based products. If you use harsher synthetic pesticides, be careful! Most of these products are highly toxic and can cause serious harm to you and consumers of your product if used improperly.

I must speak out against the use of Avid in any form on marijuana crops. This miticide is labeled "for ornamentals only" and is highly toxic. Avid is translaminar in action—meaning that it penetrates from the top of the leaf surface to the bottom as well as into the stem surface. Unscrupulous hydroponic stores which sell this under the table should be boycotted, as should growers who use it!

### PREDATOR MITES

Predator insects simply do not provide the level of control needed in the mother room. However, they do work well in a flowering room, where absolutely no spraying should take place.

There are several things to take note of when using beneficial insects. Many insecticide residues left on plant surface can adversely affect or kill beneficial insects. Also many predators need a full light spectrum, natural sunsets and long days (over 12 hours), none of which are likely to occur in your average flower room. If your supplier cannot supply you with this information, give the source company a call, they will have all of this info.

# Cannabis Culture: CHINA

## A Canadian traveller barters, smokes and climbs his way to China's highest highs. By Greg Allen

Tiger Leaping Gorge: one of the deepest canyons in the world, a serene and sensational piece of Yunnan scenery.

It was the Christmas holidays, 1998. I had two weeks off work and nothing better to do than hunt for heights and herbs in Yunnan province, People's Republic of China. The heights I knew were there; the peaks around Lijiang reach 5600 metres. As for the herbs, the Lonely Planet guide discreetly mentions their presence around Dali. So off I went on a mission, to get high in a literal/physical/spiritual sense and escape the mundane rigors of a responsible job in South-east China.

## LEAVES & SEEDS

Cannabis has a long history of cultivation in China, dating back as far as 5000-6000 years. It was grown along with millet, wheat, beans and rice in the earliest Neolithic farming communities and was regarded as one of the main crops in ancient China.

Until cotton was introduced to China about 1000 years ago, cannabis was the main cloth worn, a fact proven by both ancient texts and archeological discoveries. Pure cannabis textiles were found in tombs dating back to 1700 BC, and imprints of cannabis textiles and cordage on pottery fragments have been carbon dated to about 4000 BC.

Paper was another cannabis product long in use in China. The oldest piece of paper in the world was recovered from a tomb in Shaanxi dating to about 100 BC. A tomb in Xinjiang offered up white cannabis paper shoes sewn with white cannabis thread dating to 1100 AD.

The use of cannabis seed for food is also well documented as far back as 200 BC. It was placed as one of the "five grains" of ancient China along with barley, rice, wheat and soybeans. Cannabis remained a staple of the diet until the 10th century when other higher quality grains became widespread. Hemp seeds have often been found in storage jars inside tombs.

## MODERN CHINESE CANNABIS

Cannabis has been cultivated in nearly every province and climatic zone in China. It is still used in some areas for making rope, clothes and other textiles. The seeds are pressed for oil or eaten raw or roasted as snacks between meals (especially in northwest Yunnan). Tibetans also mix seeds in buttered tea.

Some plants in Xinjiang and Yunnan are illicitly planted for smoking

*Ban Wan Guest House: 2300 metres high and climbing.*

but cannabis smoking is not popular or widespread among Chinese.

Chinese growers realize that the female plants are used for smoking and contain more medicinal properties. In specific areas where more of the drug plants are grown (Kasgar, Hetian and Asku in Xinjiang), the upper inflorescences, younger leaves and resin gland secretions are used for making cigarettes.

On the other side of China, in Shandong province among others, there is a small but thriving modern hemp industry cultivating for fiber.

## CHINA CANNABIS TRAVELS

For those with a spirit of adventure and a desire to venture into an easily accessed but totally wild part of China, as well as a chance to score some local herb, I recommend northwest Yunnan. While the smoke won't blow you away, the natural beauty of the area surely will.

During my stay in China, I had planned ahead and brought a few of Marc Emery's seeds with me. Bringing a small handful of seeds across the border is basically a no-risk affair, so a few Arizona Big Bud plants provided me with at least a bit of leaf smoke for my mental health. But I knew there was local cannabis somewhere in Yunnan, hence my mission.

Dali, in northwest Yunnan was my first stop; a popular tourist place with rumors of herb. Dali is an interesting old town above Erhai Lake and under the Cangshan mountains, with the streets

*Eight feet of male cannabis growing outside a home in Dali.*

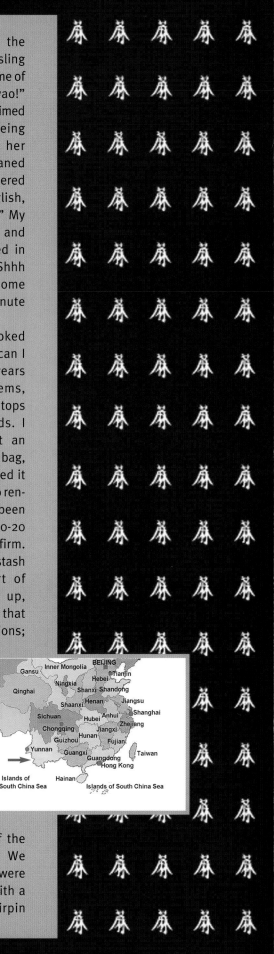

been after me on the street earlier, hassling me to purchase some of her trinkets. "Bu yao!" (don't want) I exclaimed again. Finally seeing the futility of her attempts, she leaned forward and whispered in reasonable English, "You want ganja?" My ears perked up and I quickly answered in the affirmative. "Shhh . . ." she said. "Come with me, one minute away."

The ganja looked just like the Mexican I used to buy 30 years ago; leaves, stems, occasional female tops with lots of seeds. I transferred about an ounce to a smaller bag, then nearly dropped it when she said "100 ren-mibi" ($12). I had been expecting maybe 10-20 rmb, but she was firm. As I needed good stash for the next part of my trip I bucked up, rip-off or not. A few pipefuls that evening confirmed my suspicions; harsh and low quality but there was THC in there.

offering the usual blend of markets, cafes and street hawkers.

While wandering aimlessly along a minor street, I was confronted by an 8-foot cannabis plant growing alongside the wall of a house. Alas, it was a male. Even so, I plucked a handful of ripening top leaves and dried them on the edge of a sunny window of my guesthouse. The next day I added them to the small bag of homegrown I had with me for a new blend.

Later the same day while sitting in a café nursing a big bottle of Chinese beer, my opportunity to score walked in the door. This lady had

Early the next day saw me on a local bus to Lijiang and then, the day after, on to the small town of Daju. Spectacular mountain and valley scenes presented themselves at every turn of the bumpy and dusty road. We crossed several passes that were over 3000 meters, ending with a long drop and endless hairpin turns down to the picturesque valley of Daju. Here was the eastern start of the Tiger Leaping Gorge, my next adventure.

Recently named a World Natural Heritage site, Tiger Leaping

*A high Chinese New Year on Jade Snow Dragon Mountain.*

Gorge has a well-deserved popularity with adventurous souls from around the world. It is one of the deepest canyons in the world, with the Yangtze River cascading through a narrow slot at about 1700 meters and seemingly vertical walls soaring up to snow covered peaks at 5600 meters.

After crossing the gorge in a rubber boat, a short few hours brought me to the welcome site of Wan Qu's Guest House and the promise of cold beers and a spliff. Sure enough, a bag of the proprietor's home stash and a pipe lay on the table, free for sampling. This ganja was nothing spectacular, but the real trip was lying back in the sun and soaking in the atmosphere of this serene and sensational piece of Yunnan scenery.

The next morning it was ganja pancakes for breakfast, a specialty of the house. Then I was set up for a leisurely walk, pacing the sun's progress down to the rapids, 400 meters below. I spent the next several days hiking in high and dramatic passes of the Tiger Leap Gorge.

After five fine and eventful days in the Tiger Leap Gorge, I eventually stumbled into the noise and jumble of Qiaotou, at the western edge of the gorge. The advice around was that a person could find ganja at the Peaceful Café. The last page of the café's menu had a bold and clear message: "If you want ganja, ask Xiao Hu."

While waiting for the bus back to Lijiang, I asked for a chance to dip into the big bag of the proprietor. The seedy buds looked a bit better than my previous scores, so I grabbed a generous handful. "How much?" I asked, hoping not to get ripped off like in Dali. "You say," she replied confidently. "Ten rinmabi?" I questioned. "Okay," she replied quickly so I think we were both satisfied with the transaction.

If you're traveling around East Asia and intend to dip in for a taste of China, you could (and probably would) do a lot worse than this herb-friendly part of Yunnan. The air is clean, the folks are friendly, and there is a fascinating minority culture. The food and the travel is reasonably cheap and the dramatic landscape and hiking gets you nearly as high as the ganja.

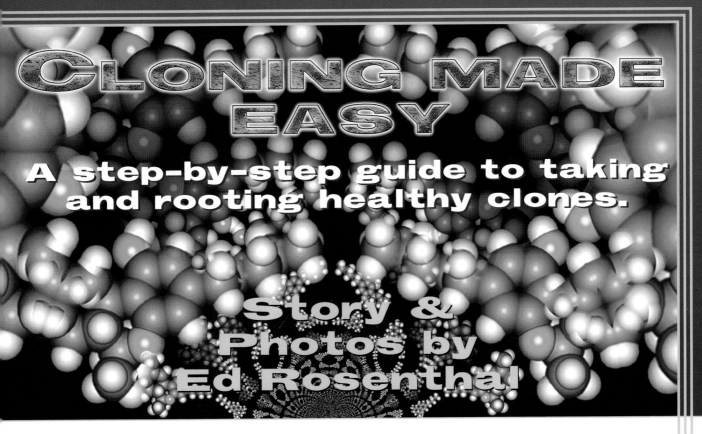

# CLONING MADE EASY

## A step-by-step guide to taking and rooting healthy clones.

### Story & Photos by Ed Rosenthal

## Clones vs Seeds

This step-by-step guide will take you through the process of taking your own clones.

There are several reasons why most growers prefer to work with clones (rooted cuttings), rather than with plants grown from seed. First is WYSIWYG – what you see is what you get. Second, clones produce a uniform garden which will mature all at once. Third, clones are easier to grow since they are all female and no males are present, ensuring unpollinated buds.

A clone is an exact genetic duplicate of the plant from which it was taken. There is no guesswork as to how it will turn out. Plants grown from seed are variable because they are the result of a toss of the genetic dice. Even plants from stable varieties such as Sensi Seeds' Skunk #1 vary quite a bit in growing habits, maturation, yield and potency. This is even more evident with some of the newer hybrids such as

a Bubbleberry x Widow or Jack Flash, both hybrid crosses. Plants grown from packets of these seeds exhibit tremendous variation.

Uniform gardens grown from clones are much easier to maintain. The plants have the same growing habits so they are easier to space and prune. They mature at the same time so there is no downtime in the garden waiting for stragglers to mature.

With seed, about 50% of the plants will become males. These plants will be discarded in order to keep the female buds unpollinated. Growing males until they are sexed is essentially wasted time and space. Any undiscovered male which sneaks open a few flowers can and will pollinate the garden. This has ruined many great crops.

Cloning is not rocket science but the plants and clones must have their needs met to grow vigorously. The most important requirement is to maintain cleanliness and hygiene around the plant and

cuttings. Cut plants are especially susceptible to infection so it is important to minimize the risk.

## Equipment

You will need fertilizer, inoculant, light, pH adjuster, planting cubes or soil, razor blade, rooting dip, scissors, sterilizing agent, trays, watering can and possibly a heat mat.

## Fertilizer

Use a balanced fertilizer. There are hundreds of brands. Three popular ones are Miracle Gro Tomato Plant Food 18-18-21, Max Sea 16-16-16 (tel 888-629-7324) or Peter's 20-20-20. These fertilizers work well in once-through systems. However, in recirculating hydroponic systems, these fertilizers are not pH stable and must be adjusted every few days and changed weekly.

Hydroponic fertilizers with A-B-C components should be adjusted to about a 1-1-1 ratio. They are usually used for about two weeks. Hormex and SuperThrive can also be added to the irrigation water.

## Inoculant

There are many bacteria and fungi which can be used as inoculants. They work in different ways. Some attack certain plant pathogens. Others form a symbiotic relationship with the roots, functioning as extensions of the root system and at the same time forming a physical barrier to pathogens. Two inoculants are used here, Rootshield (tel 800-877-9443) and PHC Biopak (tel 412-826-5488).

PHC BioPak is a bio-stimulant composed of bacterial strains which form symbiotic relationships with the developing roots and reduce susceptibility to disease.

## Light

Cuttings that are rooting do not need extremely bright light, only about 250 foot-candles. This is easily achieved using a single fluorescent light per foot of width of garden. (For example, a 2' x 4' tray would use two 4' tubes.) The ideal light for rooting is a tube high in blue light, 5000-6000 Kelvin. This spectrum encourages root growth, as well as sturdy stem and leaf growth after root initiation.

Tubes providing this light include GE Chroma 50's and Vita-Lite. Many other brands have similar tubes. If these aren't available then a "cool" white will do, but stay away from "warm" whites.

Metal halide (MH) and high pressure sodium (HPS) lamps can be used but they emit very bright light. MH lamps have a good spectrum for rooting, while an HPS will slow rooting down by a few days. A 400-watt MH lamp emits about 40,000 lumens. I have seen it used in a 10 foot square, (100 square feet) but it may be too much light with an efficient reflector. A 250 watt lamp over the same area will provide adequate support for root initiation. The lamp must be placed 6 feet above the garden in order for the light to spread out enough.

## pH Adjuster

pH up or down should be used to maintain the water at 6.2-6.5 each time the plants are irrigated.

## Planting Cubes or Soilless Mix

Oasis or rockwool cubes or a sterile vermiculite-perlite, peat- or bark-based soilless mix can be used to root the clones. Each has its advantages and uses.

• Oasis is a very porous plastic cube which is used for cuttings. It comes in many sizes, but the one used in the photographs is made of 1.5" square blocks connected on a 10" x 20" sheet. It is easy to use and is very light until dipped in water. Then its weight increases incredibly as the pores hold water. Once dipped it is ready to use.

• Rockwool is composed of tiny strands of basalt rock, manufactured by melting the stone and then spinning it in much the same way cotton candy is made. For agricultural use it is formed into blocks of various sizes as well as loose material called flocking. The 1.5" square blocks work well for cloning. In several controlled experiments, cuttings in rockwool rooted 2-3 days earlier than those in Oasis.

Rockwool cubes tend to be very alkaline and must be buffered by soaking them in a strong acid solution for a day. This neutralizes the pH. Rockwool, especially the flocking, creates dust when dry that is bad for the lungs. For this reason it should be wetted before working with it. It should be handled only with gloved hands.

• Sterile commercial peat- or bark-based mix should be used for clones destined for outdoor placement. This is the preferred rooting medium since it leaves no residue to scar the earth.

Clones can also be rooted in perlite, vermiculite, or a mixture of the two.

## Razor Blades

Razor blades should be sterile and sharp. They can be re-used several times if sterilized after each use. New blades are sterile. Single edge industrial blades are available at paint and hardware stores very inexpensively.

## Rooting Dip

Rooting dips contain synthetic variations of a plant auxin which promotes rooting. These are usually listed as NAA and IAA although there are several other variations. In addition they may contain plant or seaweed extracts as well as B1. The operation described here experimented with a number of different brands as well as a control using no dip. The result was inconclusive. The controls performed better than Rootone alone, but Hormex and Woods had slightly better results.

## Scissors

Scissors should have a sharp point, be well honed and move with little resistance. Size preferences run from 3"-9". There are a myriad number of styles.

## Sterilizing Agents
### Bleach

In the past bleach was used as the main sterilizing agent. However it has been replaced by less noxious products. Bleach creates unhealthy fumes and is harmful to the environment.

### Hydrogen Peroxide

Hydrogen peroxide is a powerful oxidizer which kills pathogens using a chemical burn. Its formula is $H_2O_2$, so it contains one more oxygen (O) atom than water. This oxygen atom jumps easily to pathogens and other particles, oxidizing them. When the oxygen atom is released, the remaining formula is $H_2O$ – water.

### Zero Tolerance

This brand-name sterilizing agent has a formula of $HO_2$, so it is even more reactive than hydrogen peroxide. The oxygen atoms react with impurities in the water, killing pathogens by oxidizing them. The molecular breakdown results in water and a release of oxygen (tel 888-273-3088).

### Physan 20

This greenhouse disinfectant has long been used in the industry. Its active ingredient is ammonium chloride, an oxidizer. It leaves no residue. Manufactured by Maril Products, Inc (tel 714-544-7711).

## Trays

Several trays are required. The cubes or containers are held in a 10" x 20" open-sided greenhouse tray. This tray is suspended 2-3 inches above a larger tray. This is accomplished by placing small plastic flower pots on the four sides of the larger tray. These containers will support a metal grill on which the tray then fits. The 10" x 20" tray should be covered with a deep fitting cover, or else a cover can be placed over the larger tray.

## Watering Can

Any standard watering can which pours water out gently will do.

## Heat Mat

The plants should be maintained at about 72-74°F (22-23°C). If the temperature in the clone room dips below this, a heat mat should be used to maintain these temperatures. Get one that has a regulator or which is pre-adjusted to 72 degrees.

## Cloning step-by-step

1. Choose the plant you wish to clone. It should be the one you find overall the best. It should be a great smoke and high, as well as maturing in a reasonable time and producing an above-average yield.

*An Afghani hybrid is chosen. Low light levels caused a bit of stem elongation.*

2. Prepare the fertilizer mix. Use a TDS meter to adjust the solution's strength to a reading of about 400 ppm. (A TDS meter measures dissolved solids.) Add hormones and SuperThrive as recommended. Adjust the pH to 6.2-6.4.

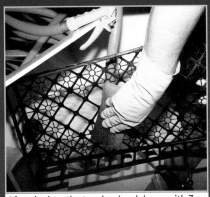

*After rinsing, the tray is wiped down with Zero Tolerance in water solution. This sterilizes everything, so the trays don't need to be thoroughly cleaned.*

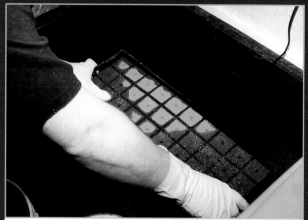

*The sheet of cubes is placed in the tray and dipped in nutrient/water solution.*

*Smaller branches are cut off the main branch.*

3. Prepare the medium and tray. If you are using Oasis, place the sheet of blocks in the tray and then either dip it in the fertilizer solution, or use the watering can to irrigate it until it is saturated and dripping.

If you are using rockwool, it should have been soaking in an acid solution for a day before this. You should then let it drain for a few minutes, and then rinse it with the fertilizer solution.

If you are using commercial mixes in small containers or a 2' x 2' tray, they should be irrigated until dripping and then allowed to drain.

You can re-use equipment, but it must be sterilized using a sterilizing agent.

Cloning rooms are favorite places for pathogens to hang out because there are a lot of vulnerable plants around sporting open wounds. Used equipment such as trays, containers, covers and tubing are infected with pathogens. They must be cleaned thoroughly with soap and hot water. This will eliminate most infectious agents, but not all of them. This is where a sterilizing agent comes in. Sterilizing agents don't clean, they kill pathogens. It doesn't matter to the plant if there's a little dirt around as long as it's sterile.

When using a sterilizing agent the cleaning doesn't have to be as thorough because the wash is not being used to sterilize.

4. Using sterile scissors, cut a large enough piece from the plant to create the number of cuttings that you need. People have many

styles of doing this. This clone operation cuts the main stem of the plant fairly severely. Then they trim the rest of the plant. They trim smaller and side branches back, but make sure to leave active growth sites so the branches can continue to grow.

5. The branches are cut up into individual 3-4" cuttings. Several cuttings can be made from a single branch. There is no minimum or maximum diameter for cuttings. They can be made from both thin shoots and main branches. Each should be topped by a healthy growing tip.

6. Each branch is trimmed. Using scissors all leaf material except the top growing tip is removed from the cutting. Make sure to trim off small growing tips below the top. Hold one where the growing tip meets the stem and then clip it off.

*Each of these is cut up so that each piece is a single cutting. All growth below the top crown is removed from the cut to limit transpiration. Any large leaves left are trimmed to a circumference of about 1" to 1.5 ".*

*Most of the greenery is cut off, leaving only the lower branches which have been in the shade.*

*The cuttings are lined up evenly at the crown, and then cut about 4 inches from the top using a sterile razor.*

The cuttings are dipped in rooting solution as suggested on the label.

A tray is placed over the clones to hold in moisture.

7. The prepared branches are lined up for the final stem trim. Rather than the stem end, all the crowns should be lined up evenly. It is usually convenient to do between 10 and 20 pieces at a time. Once they are lined up, cut them with a sterile razor.

8. Soak the clones in the rooting dip for the recommended time.

9. Push the clones into the hole in the rockwool or Oasis. Then pinch the hole shut using thumb and index finger. In commercial mixes if the stem is thin you may have to push a 1" deep hole using skewer or coffee stirrer.

10. Repeat this process until the tray is complete.

Stem is pressed into Oasis cube hole and then the hole is pinched closed using thumb and forefinger.

11. Place the 10" x 20" tray in the larger tray. It should be sitting on the grill about 3" above the bottom of the tray.

12. Irrigate using the watering can with the fertilizer solution and Rootshield or other inoculant. Water will drip into the larger tray. Leave it and add more water so that there is a one-inch level

Tray is placed in larger holding tray. A metal grill keeps it above the one and a half inch water level in the holding tray.

of water in the large tray, creating a humid environment around the sheet.

13. Place cover on tray.

14. Turn light on. The light remains on continuously.

15. Turn heat mat on.

16. The following day, water with PHC BioPak in solution.

17. Irrigate after three days using the fertilizer solution. Gently pour the water over the cubes. Water over the clones to supply water and nutrients directly to the leaves and stem. Use a watering can with a gentle spray, making sure to saturate every cube. Replace cover. Repeat this every three days so the plants are irrigated three or four times during rooting. The plants are therefore watered days 1, 2, 5, 8, 11 and 14.

Tray has two hoses. The bottom hose has a valve on it and is used to drain the tray before moving or cleaning. The second hose is the overflow valve.

18. When the roots have grown through the bottom of the cubes remove the cover. New growth will be apparent on the crown, too. The clones are ready to plant but can be kept in the tray for up to three weeks. To keep the plants healthy but slow growing, keep the fertilizer at a TDS of 400 ppm and the light intensity low. To encourage fast growth the TDS is raised to 800 ppm and the light is increased to an input of 20 watts per square foot.

Good luck and happy cloning!

Nine days after starting, the roots are quickly growing through the Oasis cubes.

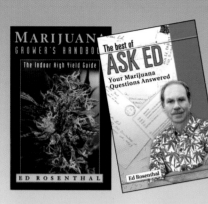
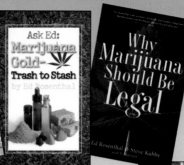